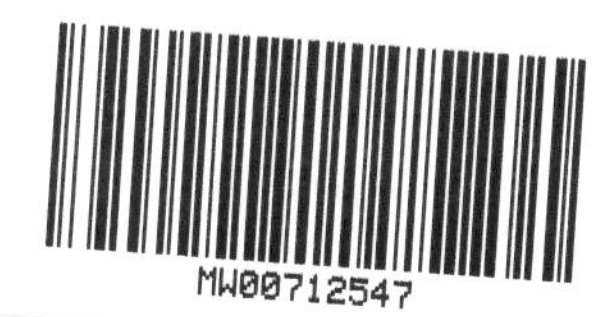

M@thPak

Enhanced with Graphing Utilities

2nd Edition Series

Michael Sullivan

Michael Sullivan III

PRENTICE HALL
Upper Saddle River, NJ 07458

Editor in Chief College Math: Sally Yagan
Associate Editor: Audra J. Walsh
Special Projects Manager: Barbara A. Murray
Media Production: Peter Silvia
Marketing Manager: Patrice Jones
Media Buyer: Tom Mangan
Cover Design: Steven Gagliostro

Upper Saddle River, NJ 07458

Printed in the United States of America

10 9 8 7 6 5 4 3 2

ISBN 0-13-040314-8

Prentice-Hall International (UK) Limited, London
Prentice-Hall of Australia Pty. Limited, Sydney
Prentice-Hall Canada Inc., Toronto
Prentice-Hall Hispanoamericana, S.A., Mexico
Prentice-Hall of India Private Limited, New Delhi
Prentice-Hall of Japan, Inc., Tokyo
Editora Prentice-Hall do Brazil, Ltda., Rio de Janeiro

Contents

If you need technical support, please call the Prentice Hall Media Support Line:

1-800-677-6337 Monday – Friday, 9 am – 4:30 pm, EST

or email us at media.support@pearsoned.com

Get it Together.

Don't you spend enough time trying to organize your life –
Without having to organize your homework?

You've got a busy life. You don't have all day to study math. Use MathPak and study smarter!

Welcome to M@thPak

a CD/web learning environment for courses in Precalculus Mathematics. Using your CD and a web connection, you will have access to MathPro Explorer tutorial software, your Student Solutions Manual, Quizzes, Tests, topical Internet links, Homework Starters, a Graphing Calculator Help manual, Animations, and a wide range of content to help you succeed in your Precalculus course.

Here's a quick view of all the features of MathPak:

QuickTime Demo Video - Will show you all the features in a video.

Multimedia MathPro Explorer 4.0 - MathPro is a tutorial package that allows you to practice unlimited problems and view multiple examples of similar problems. You can see complete solutions to any given problem, videos of the author solving typical problems, and helpful hints.

MathPak Website - Is a passcode protected website which contains the following materials:

Chapter Level:

Preparing for this Chapter

Using the same pedagogy as the text, this optional on-line multiple-choice test allows you to test your previously learned skills that will be called on again in this chapter. Self-grading, the test will alert you to any areas where you may need refreshing along with the page references to the text where you can find the material to review. This feature is mirrored in the test on the first page of every chapter (excluding Chapter 1).

Chapter Quiz

The Chapter Quiz offers you the chance at trying your hand at typical problems covering the entire chapter. Page references are offered when you find yourself needing a little extra help. You can email the results of your quiz to the instructor easily through one click of a button. In the textbook, you can select the blue numbered problems from the Chapter Review Exercises to give yourself a sample chapter quiz/test.

Chapter Test

Similar to the Chapter Quiz in content, the Chapter Test does not allow for hints or suggestions. The test is automatically graded and can be sent to the instructor via email.

Section Level:

True/False Reading Test

Do you really know what you're reading for when you read a math book? This reading quiz is written by the author and will give you a good idea of what the key concepts are of the section. It is self-grading and suggests the page numbers for you to go back and read. This too, can be submitted to the instructor if you wish.

MathPro Section Objectives

This section reminds you of the mathematical objectives covered within the section and suggests the use of MathPro Explorer 4 to try your hands at successfully mastering those objectives.

Features that are always available to the student:

MathPro Explorer 4

MathPro Explorer 4 will have a navigational button that will launch the full student version at anytime. You will not have to navigate down to the section objective level to use this state of the art software tutorial.

Solutions Manuals

The complete student solutions manuals will be available via chapter downloads from the website

Graphing Calculator Manuals

Interactive graphing calculator manuals for the TI-82, 83, 85, 86, 89, 92, HP486, Casio CFX-9850GaPlus, SharpE-9600 will be available for you via the website.

M@thPak System Requirements:

- Pentium 200-MHz processor-based computer or better
- 32MB of random access memory (RAM)
- 64MB or more recommended
- 30MB free hard disk space or network drive space for program
- Microsoft Windows 95, 98, 4.0, or 2000 OR
- Macintosh OS 8.x or OS 9 (OS 8.6 or OS 9 recommended)
- CD-ROM drive
- a graphics adapter card (VGA, Super VGA, or other Windows-compatible card) capable of displaying at least 256 colors at 640X 480 pixel resolution
- Sound card
- Quick Time 4.0
- Internet Browser 4.0 or higher
- Internet Access 28.8k or better

Easy steps to Install M@thPak

For Windows:

1. Remove MathPak CD-ROM from the back of this package.
2. Insert into CD-ROM disk drive.
3. Select and double click on the “My Computer” icon on your main screen.
4. Select and double click on the CD-ROM drive icon.
5. Open and read the “ReadMe.txt” file
6. ***If you have QuickTime installed on your computer, skip to 7.***
   6a. Select and double click on the “QT4 Insta” file. Follow onscreen directions.
7. ***If you have Netscape installed on your computer, skip to 8.***
   7.a Select and double click on the “Netscape Install” file. Follow onscreen directions.
8. Install MathPro Explorer 4.0 by selecting the “MPE4 Home” directory and double click to open the directory.
9. Double click on the “Setup.exe” file and then follow onscreen directions.
10. Open MathPak Main screen by double clicking on “MathPak.exe” file.
11. It is recommended that you view the Demo before getting started.

For Macintosh:

1. Remove MathPak CD-ROM from the back of this package.
2. Insert into CD-ROM disk drive.
3. Select and double click on the “MathPak” icon on your main screen.

4. Open and read the "ReadMe.txt" file
5. ***If you have QuickTime installed on your computer, skip to 6.***
   5a. Select and double click on the "QuickTime 4 Install" file. Follow onscreen directions.
6. ***If you have Netscape installed on your computer, skip to 7.***
   6.a Select and double click on the "Netscape Install" file. Follow onscreen directions.
7. Install MathPro Explorer 4.0 by selecting the "MPE4 Home Install" directory and double click to open the directory.
8. Double click on the "Setup.exe" file and then follow onscreen directions.
9. Open MathPak Main screen by double clicking on "MathPak.exe" file.
10. It is recommended that you view the Demo before getting started.

To enter the restricted M@thPak website:

1. Select the "Register Here" button.
2. Enter your pre-assigned Activation ID and Password, exactly as they appear on the sticker found on the inside back cover of this package.
3. Select "Validate Membership"
4. Complete the online registration form and select your own personal User ID and Password. Once your personal User ID and Password are confirmed, PLEASE WRITE THEM DOWN AND KEEP THEM IN A SAFE PLACE. You will need to use these on all future visits to M@thPak online.

If you did not purchase this product new and in a shrink-wrapped package, this Activation ID and Password may no longer be valid. However, if your instructor is recommending or requiring use of M@thPak you can purchase a new copy of this Sullivan Media Companion online or at your local bookstore. Your Activation ID and Password will be valid for one year after your initial registration.

What is the difference between the website described here and the website that is found at www.prenhall.com/sullivan?

The website currently available at www.prenhall.com/sullivan will continue to be free and accessible to anyone. It will remain basically as it is now. "M@thPak" website is accessible only when launched from the **M@thPak** CDROM. All the extra benefits and features are available only to those purchasing the **M@thPak** package.

Welcome to MathPro Explorer 4.0!

MathPro Explorer is a computerized version of your math textbook and is organized in an identical format. For each section and objective in your text, there is a corresponding section in MathPro Explorer.

MathPro Explorer generates a set of problems to solve for each learning objective. Unlike textbooks, which only show a few examples in each section, MathPro Explorer allows features that a textbook cannot:

1) View multiple examples of similar problems.
2) View the complete solution to any problem.
3) Work through an interactive step by step process on how to solve a particular problem.

MathPro Explorer can be purchased in two versions:

1) The School/Network Version is available for installation in multi-user environments either via a LAN or installed on shared lab computers.
2) The Student/Home Version is available for installation on an individual computer for an individual user.

Student/Home Version

The Student/Home Version of MathPro Explorer provides the complete functionality of the MathPro Explorer Tutorial product. This includes the following, which are structured according to the Chapter, Section, and Objective content of the textbook:

- Problem Generation
- View and/or Watch Example
- Step by Step
- View Solution
- Redo Problem
- Check Answer
- Score Summary
- Explorations

The Student/Home Version also provides the capability to generate and take Practice Tests through the use of the separate Administrator Program. The Administrator Program is described later in this document.

Explorations

The Explorations element adds Exploratory Labs that employ the following additional tools, each with extensive capabilities:

- Algebra Tools
  - Symbolic Editor
  - Assignment Editor
  - Algebra Tiles - Manipulative
- Coordinate Graph Tool
- Geometry Tool
- Spreadsheet Tool
- Data Series Graphs
  - Bar Graph
  - Line Graph
  - Pie Graph
  - Histogram Graph
  - Box and Whisker Graph
- Probability Tools
  - Spinner
  - Number Objects
- Text Tool

In order to access Explorations, the user **must** log in the first time they use the program. If the user does not log in, Explorations will not be available.

MathPro Explorer 4.0 System Requirements

WINDOWS REQUIREMENTS

- 80486/66 MHz processor minimum
- 12MB of random access memory (RAM);
- 16MB or more recommended
- 30MB free hard disk space or network drive space for program
- Microsoft Windows 95 or higher
- CD-ROM drive
- A graphics adapter card (VGA, Super VGA, or other Windows-compatible card) capable of displaying at least 256 colors at 640 X 480 pixel resolution
- Sound card
- Quick Time 4.0 or higher
- For School/Network Version, if your book contains Example Videos, additional space on the network (approximately 180-200 MB) for multimedia files

MACINTOSH REQUIREMENTS

- Power Macintosh or higher
- 12MB of random access memory (RAM);
- 16MB or more recommended
- 45MB free hard disk space or network drive space for program
- Macintosh System 8.6 or higher for the server, 8.0 or higher for the client machines, both with Virtual Memory 'On'
- CD-ROM drive
- Color Display
- Quick Time 4.0 or higher
- For School/Network Version, if your book contains Example Videos, additional space on the network (approximately 180-200 MB) for multimedia files

Microsoft Windows Home Install Instructions

Installing MathPro Explorer 4.0 upgrades all existing MathPro Explorer 3.0 books *in the install location.*

1. To install MathPro Explorer 4.0, run setup.exe on the installation CD and follow the directions on screen.
2. Choose a destination location. To install to a directory other than the default, click the Browse button.

    IMPORTANT: It is strongly recommended that you install all MathPro Explorer books to the same location. To change your MathPro Explorer 4.0 install location, choose Back and change the Destination Folder. The default location is C:\Program Files\MathPro Explorer

3. Specify the Program Folder.

    IMPORTANT: It is strongly recommended that you do NOT change the Program Folder.

4. To confirm selections choose the Next button. To change your selections choose the Back button.
5. If you have MathPro Explorer 3.0 books in the same install location, they will be upgraded automatically during the installation process. If you have MathPro Explorer books in different locations, they will not be upgraded to MathPro Explorer 4.0.
6. Once MathPro Explorer 4.0 has been successfully installed, click Finish to complete the installation process.

Upgrading MathPro Explorer 3.0 Books

1. The MPE Converter runs automatically during the installation process.

    NOTE: Potential bad data may cause problems in the converter. If the converter displays an error, simply re-run the converter. It can be found in the new Start Menu Program Folder.

    IMPORTANT: MathPro Explorer 3.0 books installed in locations other than the current install location will NOT be upgraded to run with MathPro Explorer 4.0.

2. A new Administrator will be created the first time you run MathPro Administrator after the conversion. You will need to recreate the administrator password. There is no default, so you may choose any password you would like.

3. Open Convert.log in the current install location to view student name changes (See important note below).

    IMPORTANT:
    If you logged in with only a first name (e.g., Jane) in the MathPro Explorer 3.0 book, then the program will use "Unknown" as the last name (e.g., Jane Unknown) when logging into the MathPro Explorer 4.0 upgraded version.

    If you logged in with only a last name (e.g., Smith) in the MathPro Explorer 3.0 book, then the program will use "Unknown" as the first name (e.g., Unknown Smith) when logging into the MathPro Explorer 4.0 upgraded version.

    If you logged in as Visitor, the program will use "MathPro Student" as your first and last names when logging into the MathPro Explorer 4.0 upgraded version.

Access To Multimedia Features of MathPro Explorer 4.0

MathPro Explorer 4.0 requires QuickTime to play introductory and instructional videos. If you do not have QuickTime 4.0 or higher on your machine, you must install QuickTime 4.0 after you complete the installation of MathPro Explorer 4.0.

QuickTime 4.0 is included on the MathPro Explorer CD. You can also go to

http://www.apple.com/quicktime

to download it.

Access To MathPro Explorer

You will have a new folder on your local drive called "MathPro Explorer." Each book installed will be within the MathPro Explorer directory. Click the Start button and chose Programs to access the Program Folder for the desired book. You may run the MathPro Explorer, the MathPro Help, the Glossary, or the Administrator from this group. Access to the Explorations is available from within MathPro Explorer. Please refer to appropriate sections of this MathPro Explorer User Manual for program functionality.

Uninstall Instructions

To uninstall MathPro Explorer, open the Control Panel and double-click on Add/Remove Programs. Select the appropriate author/title/version of the program to uninstall from the list of installed programs. When MathPro Explorer is uninstalled, all files that were created by MathPro Explorer will not be removed automatically by the uninstall program. You should check the MathPro Explorer or MPE directory and the book-specific directory to remove any unwanted files and/or directories after the uninstall.

For the uninstall, the program will check for shared files. If you have other books installed, the program does not remove the shared files because the other books use these files. If there is a question about a file that the program cannot answer, you will be prompted to decide whether to remove a shared file. If the book you are uninstalling is the only one installed, or if you are uninstalling all books, then you can safely remove shared files without risk of deleting files used by other programs. **Otherwise, do not remove shared files.**

Macintosh Home Install Instructions

Installing MathPro Explorer 4.0 upgrades all existing MathPro Explorer 3.0 books in the install location.

1. Insert MathPro Explorer Home CD in the CD drive.
2. Double-click the MPE Home Install CD icon to open the folder.
3. Double-click the MPE Home Install icon in the MPE Home Install window. Follow the installation instructions on the screen.
4. Choose an install location. Use the box in the lower left corner to select a different location. DO NOT select a location which contains a period (.) in the path name. For example, *Mac HD:MyNetworkPath:MathPro.Explorer* is NOT a valid path.

    > IMPORTANT: MathPro Explorer 3.0 books installed in locations other than the current install location will NOT be upgraded to run with MathPro Explorer 4.0.

    It is strongly recommended that you install all MathPro Explorer books to the same location.

    If you currently have MathPro Explorer 3.0 books in more than one location, then you must run a MathPro Explorer 4.0 Upgrade Install to upgrade 3.0 books that are not in the current install location.

5. Click the Install button in the lower right corner.
6. Click the Quit button when the installation is complete.

Upgrading MathPro Explorer 3.0 Books

1. The MPE Converter runs automatically during the installation process.

    > IMPORTANT: MathPro Explorer 3.0 books installed in locations other than the current install location will NOT be upgraded to run with MathPro Explorer 4.0.

2. A new Administrator will be created the first time you run MathPro Administrator after the conversion. You will need to recreate the administrator password. There is no default, so you may choose any password you would like.

3. Open Convert.log in the current install location to view student name changes (See important note below).

   IMPORTANT:
   If you logged in with only a first name (e.g., Jane) in the MathPro Explorer 3.0 book, then the program will use "Unknown" as the last name (e.g., Jane Unknown) when logging into the MathPro Explorer 4.0 upgraded version.

   If you logged in with only a last name (e.g., Smith) in the MathPro Explorer 3.0 book, then the program will use "Unknown" as the first name (e.g., Unknown Smith) when logging into the MathPro Explorer 4.0 upgraded version.

   If you logged in as Visitor, the program will use "MathPro Student" as your first and last names when logging into the MathPro Explorer 4.0 upgraded version.

Access To Multimedia Features of MathPro Explorer 4.0

MathPro Explorer 4.0 requires QuickTime to play introductory and instructional videos. If you do not have QuickTime 4.0 or higher on your machine, you must install QuickTime 4.0 after you complete the installation of MathPro Explorer 4.0.

QuickTime 4.0 is included on the MathPro Explorer CD. You can also go to

http://www.apple.com/quicktime

to download it.

Access To MathPro Explorer

You will have a new folder on your local drive called "MathPro Explorer Home Version." Each book installed will be within a MathPro Explorer directory. You may open (double-click) the MathPro Explorer folder to see the program icons. From here, you may double-click and run the MathPro tutorial program, the MathPro Help, the MathPro Administration, the MathPro Administration Help, or the Glossary. Access to the Explorations, MathPro Explorer Toolkit, and MathPro Explorer Toolkit Help is available from within MathPro. Please refer to appropriate sections of this MathPro Explorer User Manual for program functionality.

MathPro Explorer 4.0 Release Notes

MathPro Explorer 4.0 Multimedia Features

MathPro Explorer 4.0 now offers videos featuring the book's author to go along with selected objectives. These videos are available through the new Media Browser (previously the Explorations Browser) and directly from the Problem Sets screen. Please refer to the following explanations to see how you can take advantage of MathPro Explorer 4.0 Multimedia.

Macintosh and Windows

If you are running the **MathPro Explorer Home** version the videos will be run from your MathPro CD. Please make sure the CD is loaded when watching videos.

If you are running the **MathPro Explorer School** version the videos will be installed to the book's media directory on the network and copied to the client machine only as needed by the user.

Windows

- Videos are copied to the client machine into
  <<Windows Directory>>\mpTemp\<<Book Name>>

  Example:
  C:\Windows\mpTemp\sulalg

- This client directory will not exceed 30M or 20% of free hard drive space.
- If copying a new video will exceed 30M or 20% free hard drive space then the *least* recently accessed video will be deleted from the client machine.

Macintosh

- Videos are copied to the client machine into
  <<System Folder>>:<<Book Name>>

  Example:
  Mac Hard Drive: System Folder: sulalg

Quick Start

Follow these quick, easy steps to start using MathPro Explorer immediately.

1. Login and click the **OK** button.
2. Click the **Chapter** you want to work in.
3. Click the **Section** you want to work in.
4. Click the **Objective** from which you want to work problems. Warm-up problems or Exercise problems generate automatically. If you are working Warm-up problems, your answers will not be scored. If you are working Exercises, your answers will be scored.
5. Click the problem you want to work.
6. Read the instructions and enter your answer in the space provided.

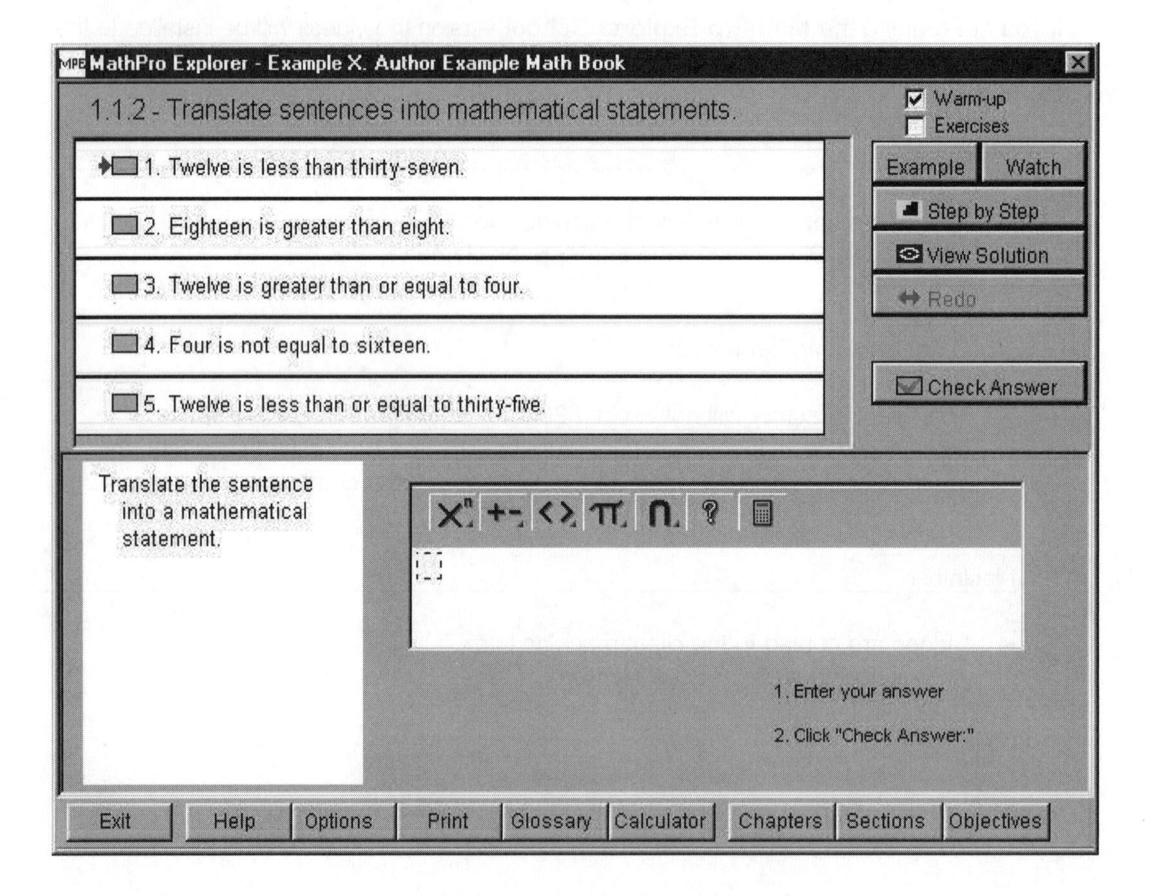

The Login Screen

The Title screen displays a graphic picture of your textbook cover. It will display briefly and be followed by the Login screen.

This screen is where you will log in the first time you run MathPro Explorer. **You will only be allowed to enter your student information once**, so make sure you enter the same username you use at school. On subsequent logins, you will automatically be logged in with you username. This allows MathPro Explorer to track your progress and your results throughout the program.

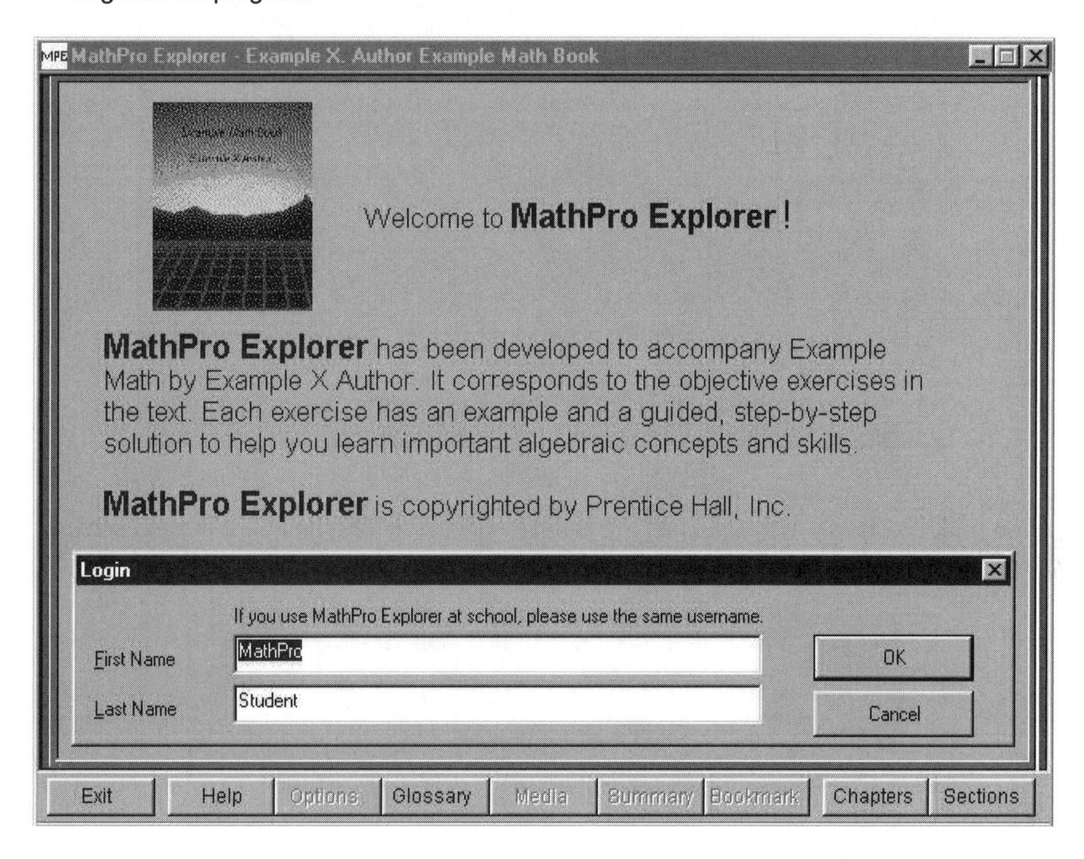

After you complete your login, you will see a dialog box that will give you the opportunity to play an Introduction Video to MathPro Explorer. From this dialog box you can also launch to the Prentice Hall website for your book. You can choose to suppress this screen for future logins. If you choose to suppress this dialog box, you can access the Introduction Video and the launch to the website from the Chapter screen.

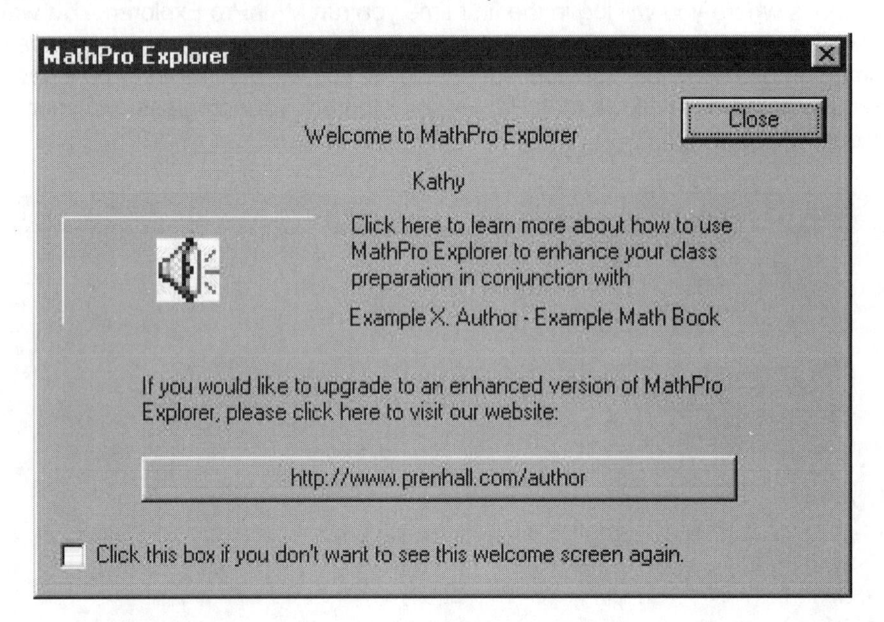

If you have previously logged in, and if you have suppressed the default Welcome screen, you will see the Welcome Back dialog box.

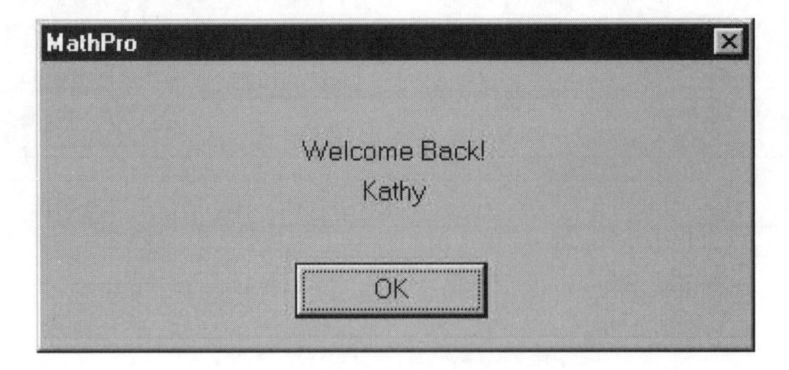

The Chapter screen displays next.

The Explorer Chapter Screen

A Course Introduction button and a Mailbox button are in the top right corner of the Chapter screen. The Course Introduction button allows you to play the Introduction Video or launch to the Prentice Hall website for your book.

The Mailbox button allows you to view messages sent by the administrator/instructor. You have the option to reply, and your messages are automatically sent to the administrator.

Each Chapter in the Chapter screen is "hotlinked," taking you to any Chapter you click.

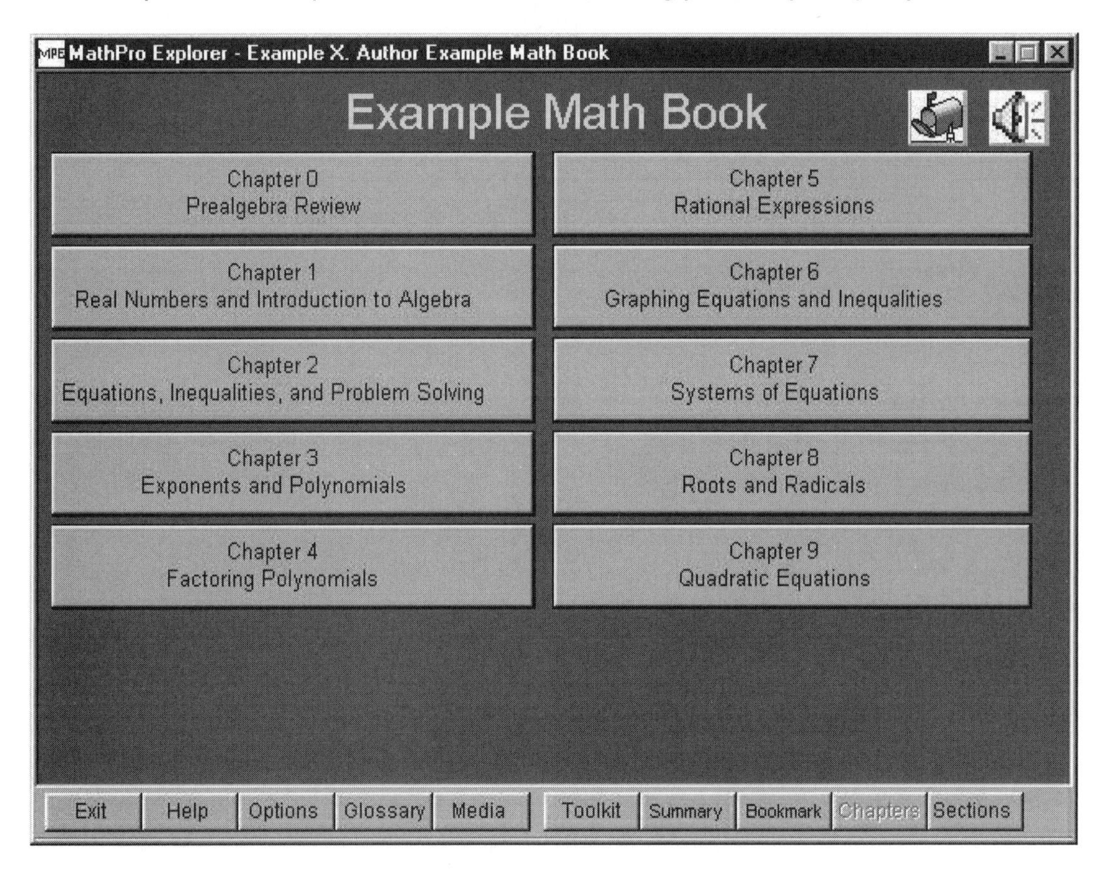

At the bottom of the screen are several buttons. If your book contains Example Videos, your screen will have a Media button. These buttons allow you to:

- Exit the program
- Access Help
- Set Options
- Access the Glossary
- View the Media Browser in which you may now also view Explorations

- Access the Toolkit
- See a Summary of your scores
- Set a Bookmark
- Go to a Chapter or Section Screen

Also, see the Explorer Button Definitions section of this document for more detail.

The Explorer Section Screen

The Section screen displays all the sections within a chapter. Clicking a Section button will take you to an Objective screen that shows the objectives for each section.

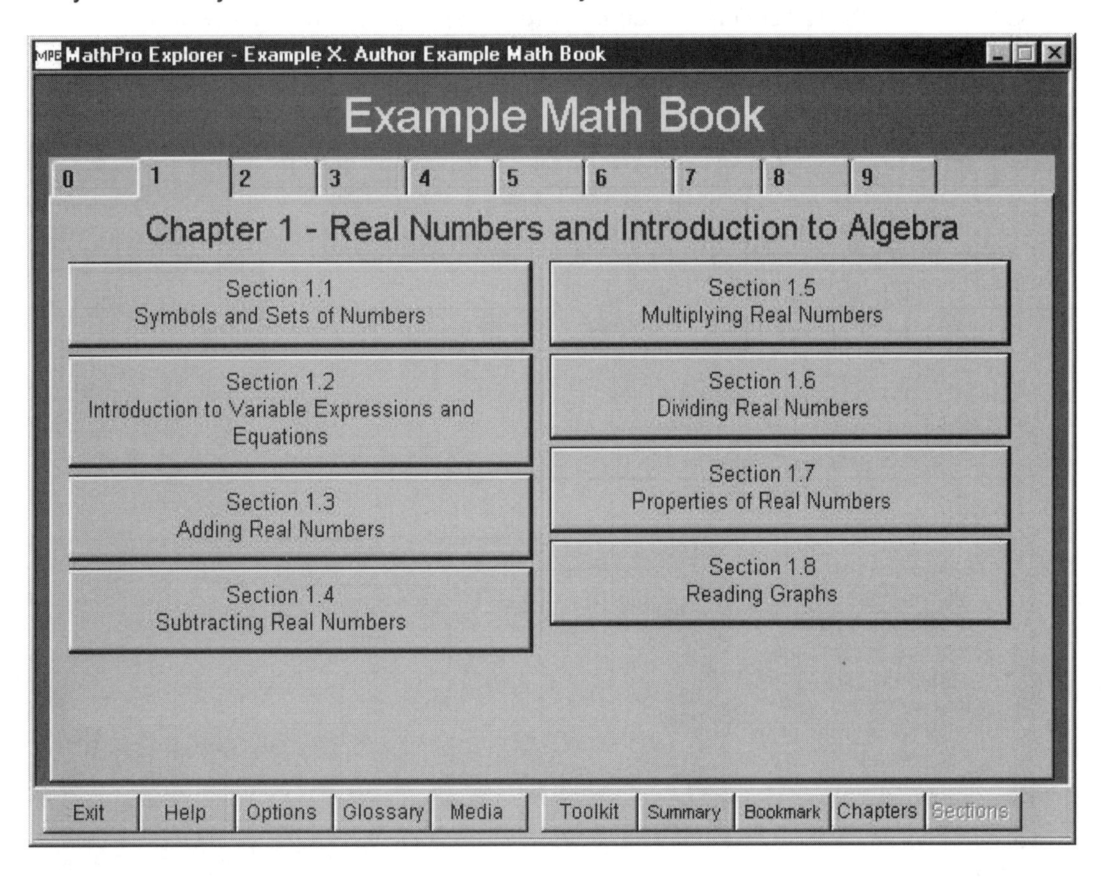

Practice Test If the administrator has created a Practice Test for a particular chapter, you will see a Practice Test button to the left of the Chapter title. Clicking this button will open the Practice Test.

The Practice Test Screen

The Practice Test screen displays the practice test created by your administrator/ instructor. After answering a problem, click the Next Problem button or press Enter. When you have finished the test, click the Submit Test button. Before you submit your test, you will have the opportunity to go back to any problem to change your answer by clicking on the problem. In the upper right corner, the screen displays how many times you have taken this Chapter Test, as well as the maximum number of times it can be taken if that number is constrained by the administrator.

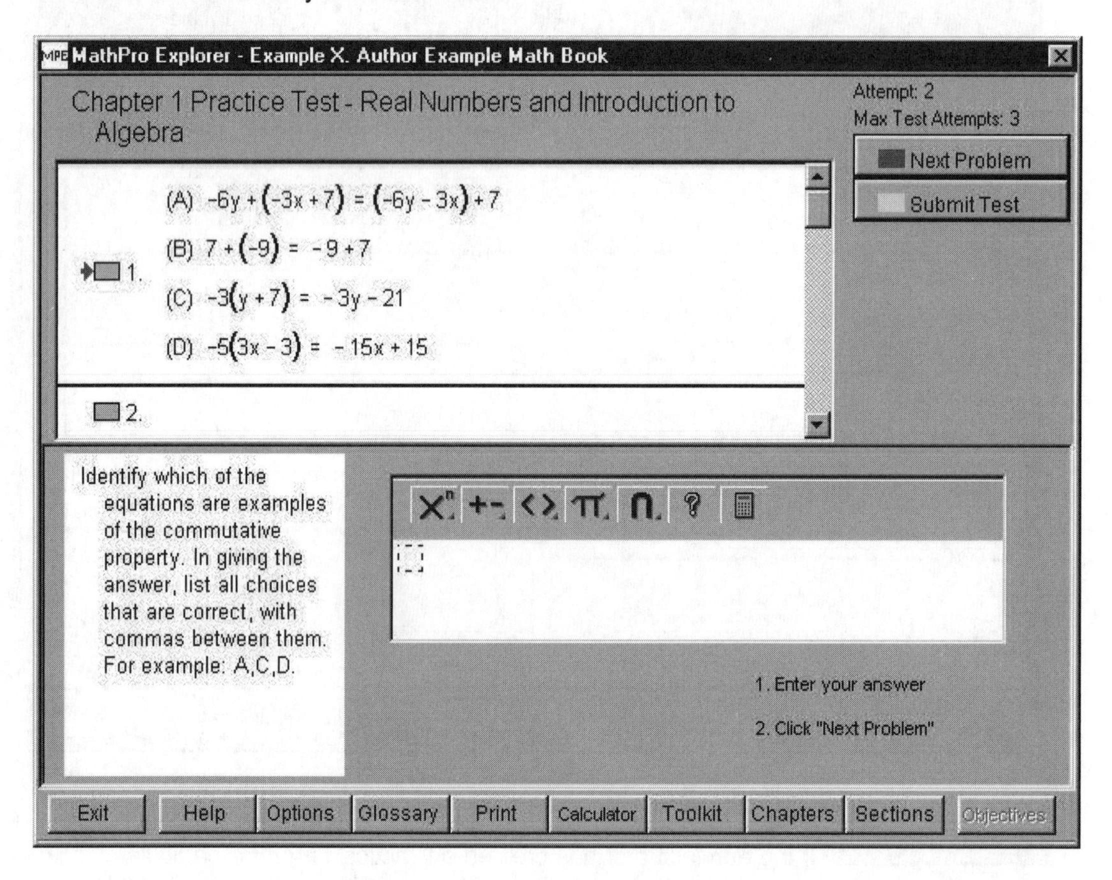

Notice that a Print button and a Calculator button are available at the bottom of the Practice Test screen.

The Explorer Objective Screen

The Objective screen displays the Objective titles for each section in your book.

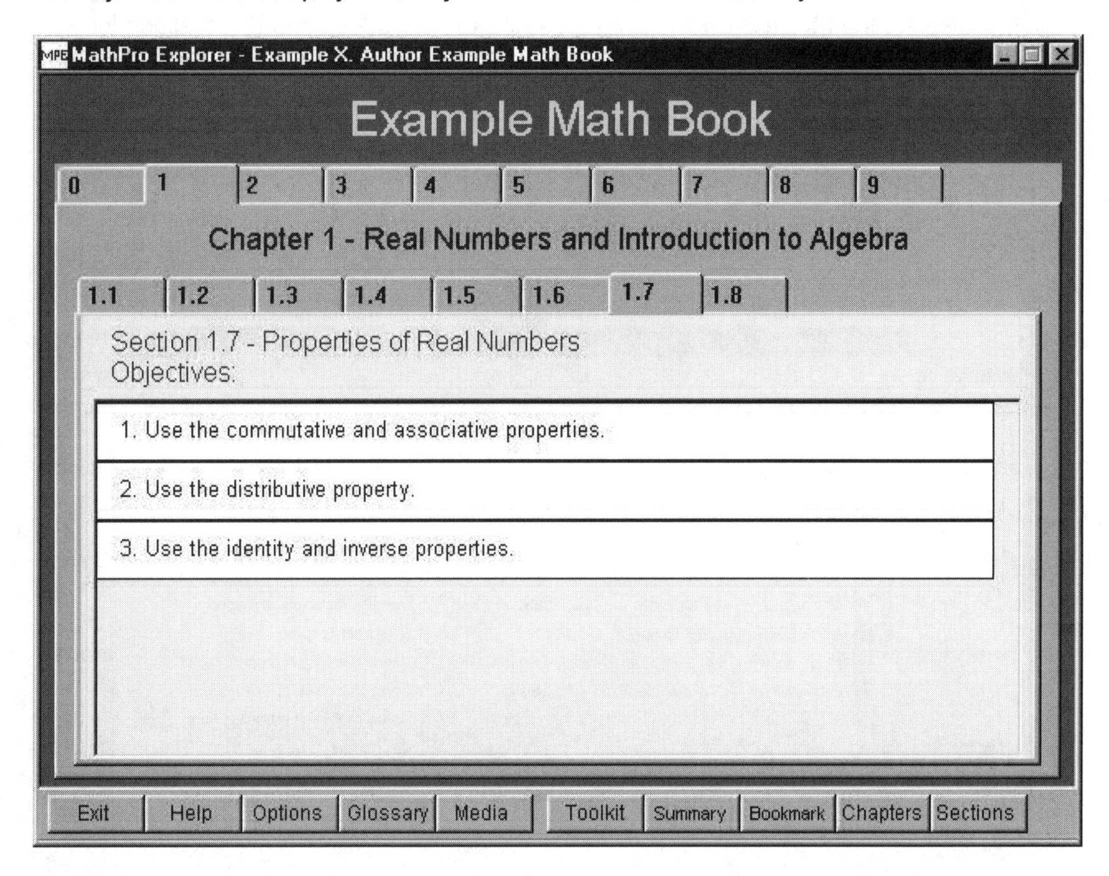

The objectives for each section correlate to those found in the textbook. To generate problems, click the Objective that you are currently studying or would like to practice. The program automatically generates 5 Warm-up problems or 10 Exercise problems each time you click, unless you have specified a different number using Options.

The Explorer Media Browser

If your book contains Example Videos, your screen will have a Media button. The Media button displays the Media Browser, which lists all example videos and Explorations for your book. You can toggle between the Explorations and the Video listings by clicking on a tab. Each tab contains a list of all explorations or videos. You can select and open an exploration or a video from the Browser at any time.

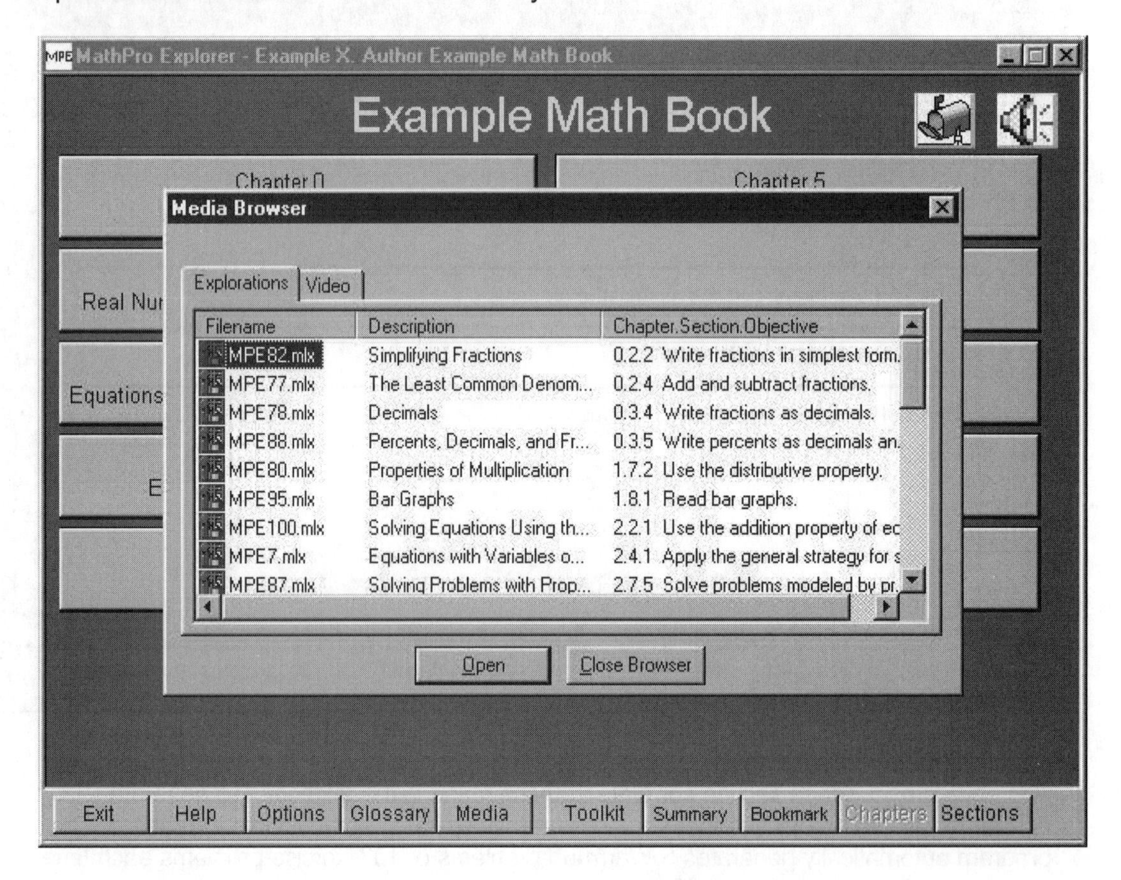

The Explorer Problem Sets Screen

When you click an objective in the Objective screen, a set of Warm-up problems generates automatically. Notice that the Warm-up box appears checked in the upper right corner of the Problem Sets screen.

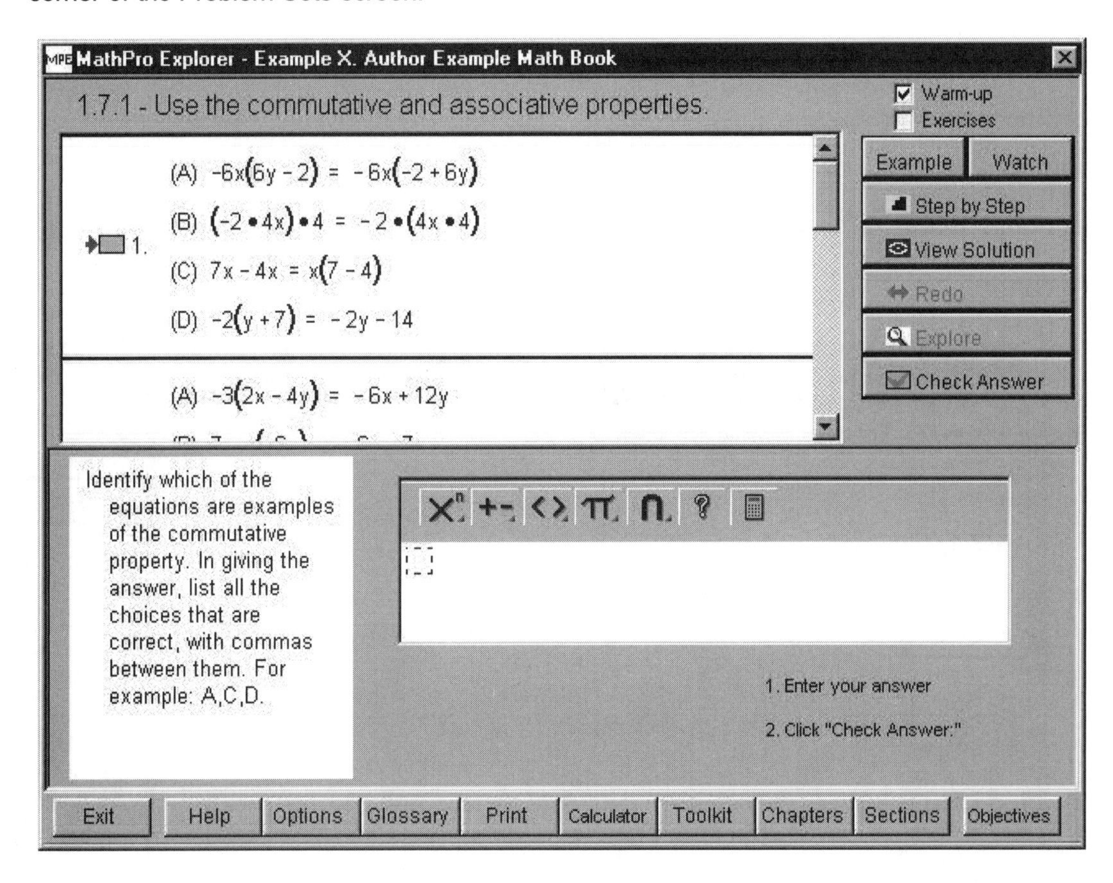

If you want Exercise problems, click on the box to the left of Exercises. Your scores are collected on Exercise problems only. Click the Warm-up or Exercise box again to generate a new set of problems of whichever type you choose.

Click the problem you want to work. A red arrow points to the selected problem.

Instructions are in the lower left window for each problem. Read them before entering any solution, as instructions may change depending on the problem you select. The answer box appears under the set of problems. There are a variety of different answer box displays that you will see and use throughout the product.

Read the instructions on the screen, follow them carefully, and enter your answer. Then check your answer by clicking the Check Answer button or hitting the Enter key.

The Print button in the toolbar at the bottom of the Problem Sets screen allows you to print out problem and status information for the current set.

Check Answer

Check Answer Once you have entered an answer in the given answer display, click Check Answer or press the Enter key on your keyboard to see if your answer is correct.

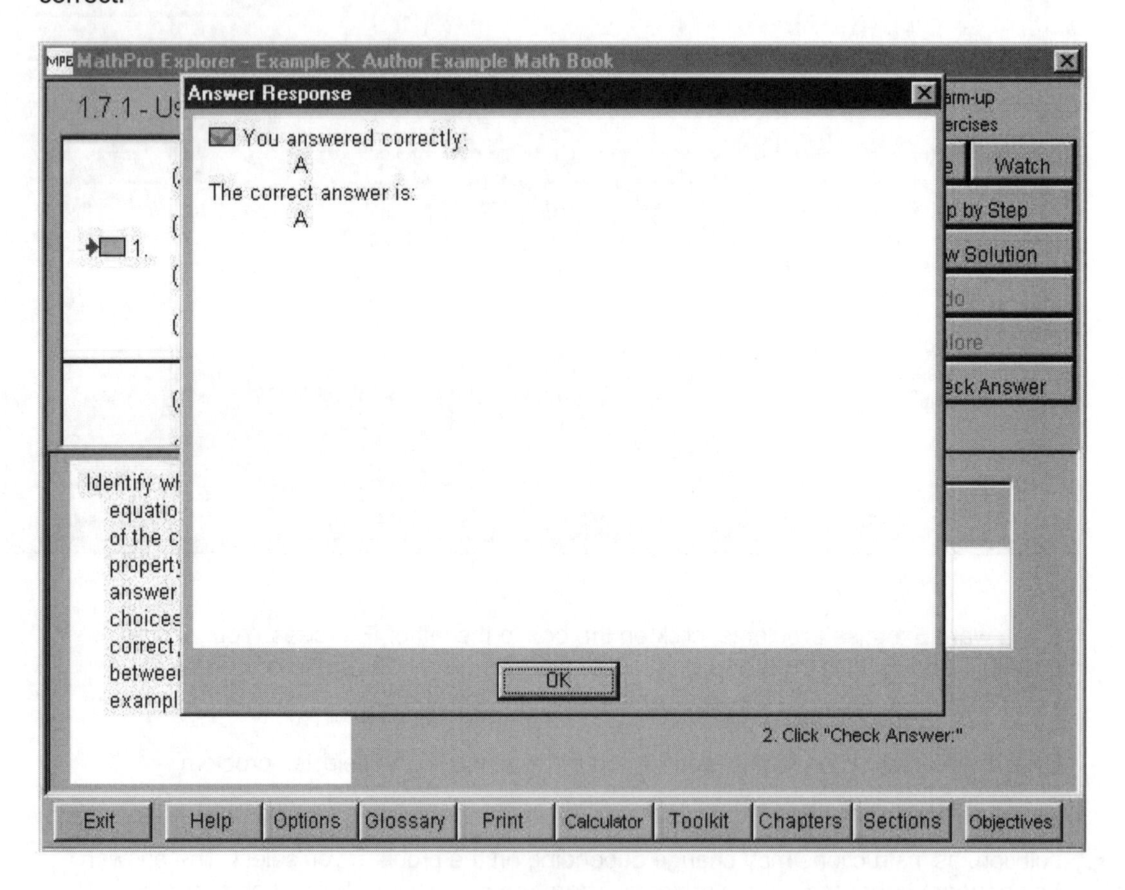

View Example or Example/Watch

View Example or Example Watch When you have selected a problem, click the View Example or Example button to see the solution to a problem that is similar to the one you have selected.

If your book contains Example Videos you will also have a Watch button available. Click the Watch button to view a video related to the current objective.

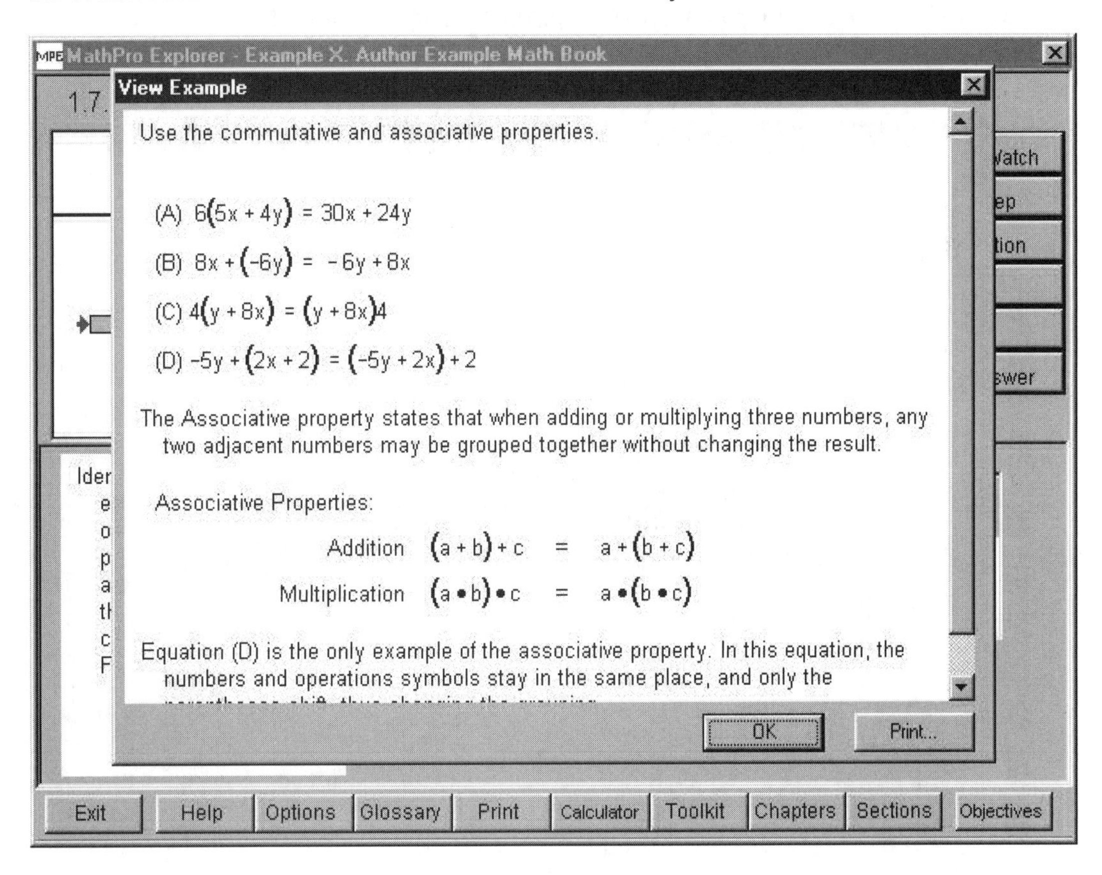

Step by Step

Click this button if you have difficulty solving the current problem. The Step by Step utility guides you through the process of solving any problem one step at a time. Use it to pinpoint the steps where you need help. Use it as often as you like within an objective until you have mastered the process for solving that particular kind of problem.

You may use Step by Step only once per problem. If you use the Step by Step utility, you will not receive credit for the problem.

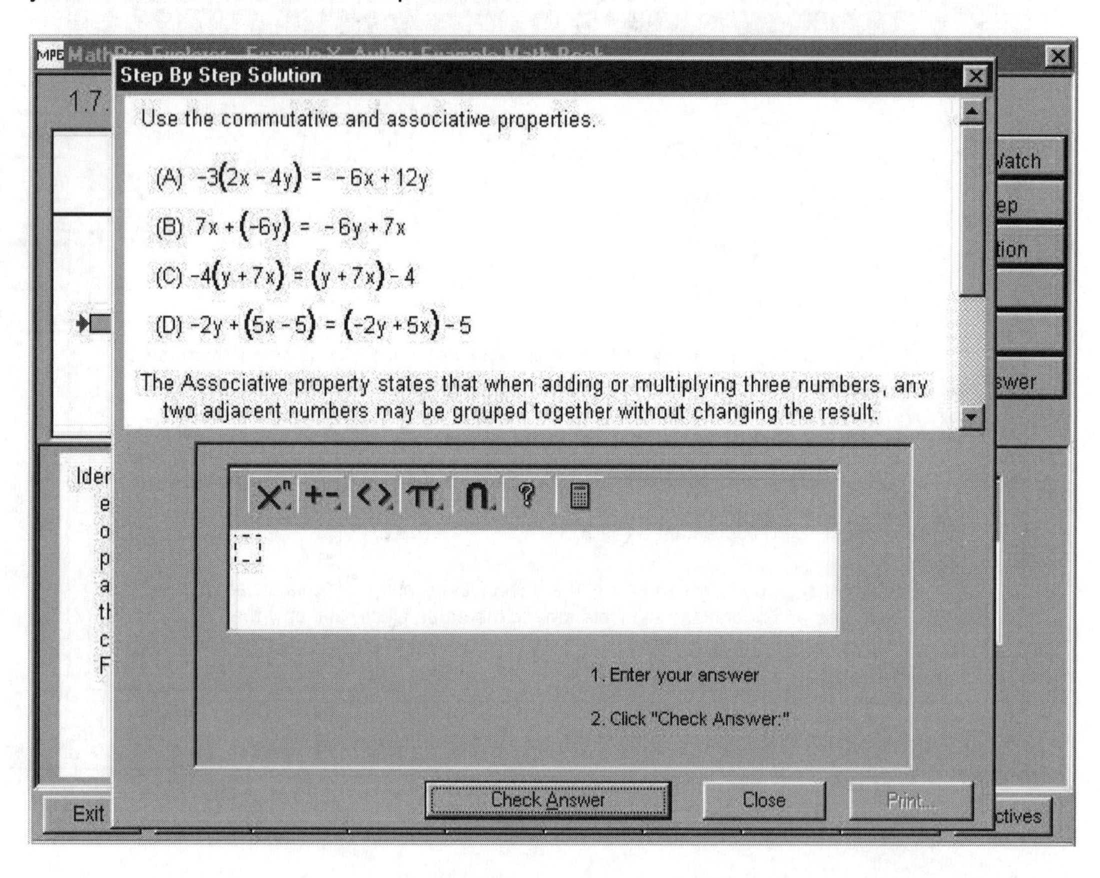

View Solution

View Solution Click this button to view the solution to the current problem. If you use the View Solution utility, you will not receive credit for the problem.

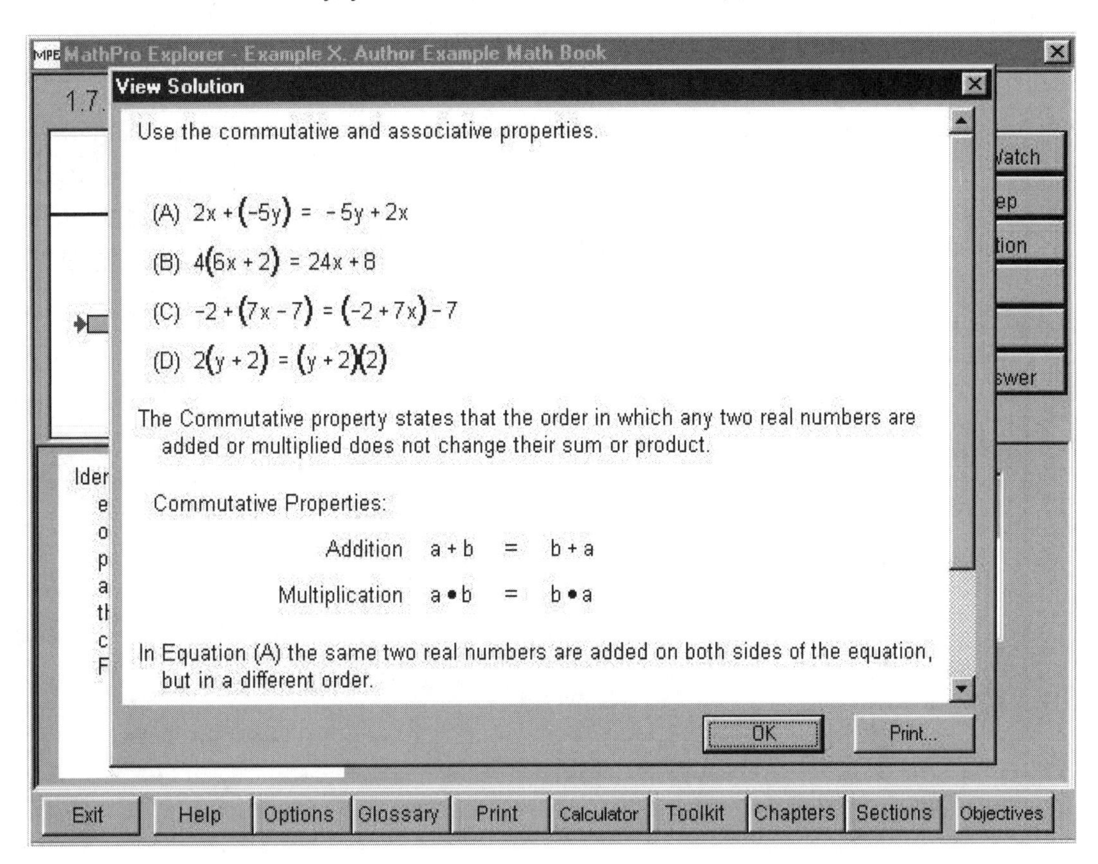

Redo

Redo If you click Check Answer and your answer is incorrect, you have the option to go back and work the same type of problem again. Simply click on the incorrectly answered problem and click the Redo button. A new problem is generated. The system will track both the incorrect problem and the new problem, and the scores are added to the cumulative scores for the objective if you are working Exercise problems.

Explore

If an Exploration accompanies the current objective, the Explore button is active. Open an Exploration by clicking the active button.

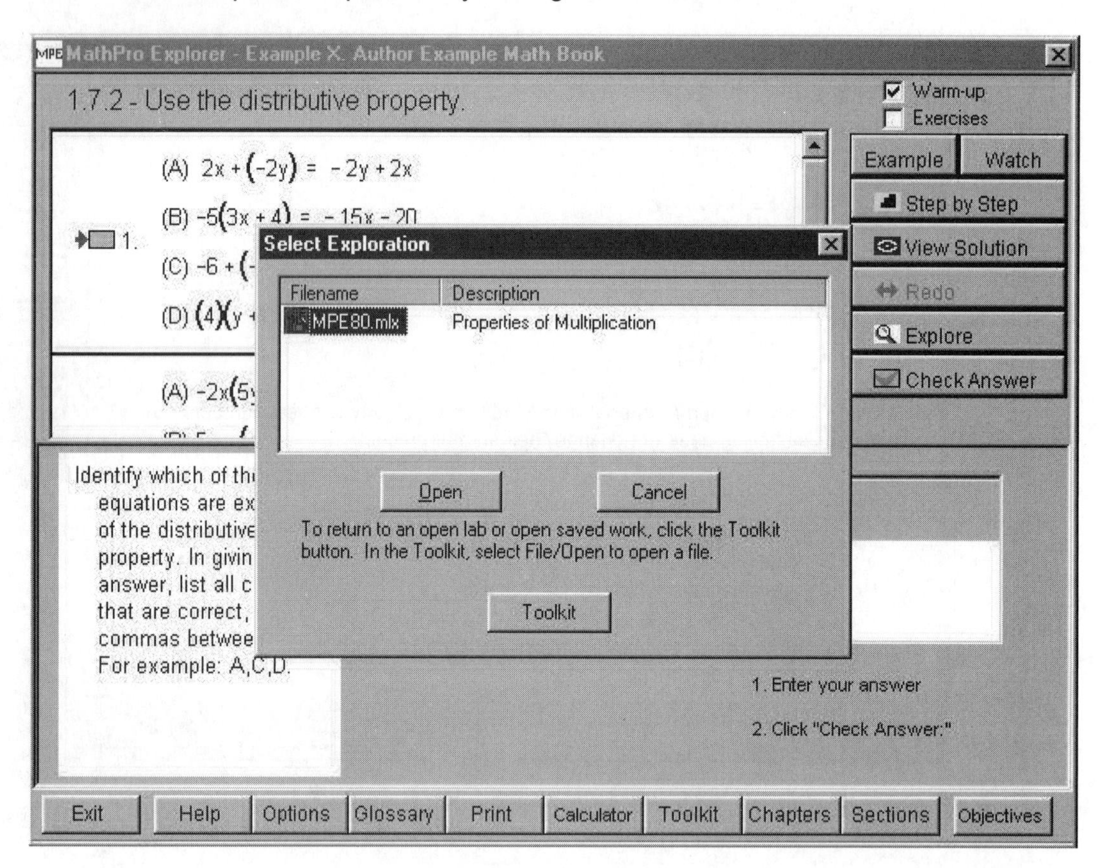

If there are no Explorations associated with the objective, the Explore button is inactive. If you click on the inactive Explore button you will see the following dialog box:

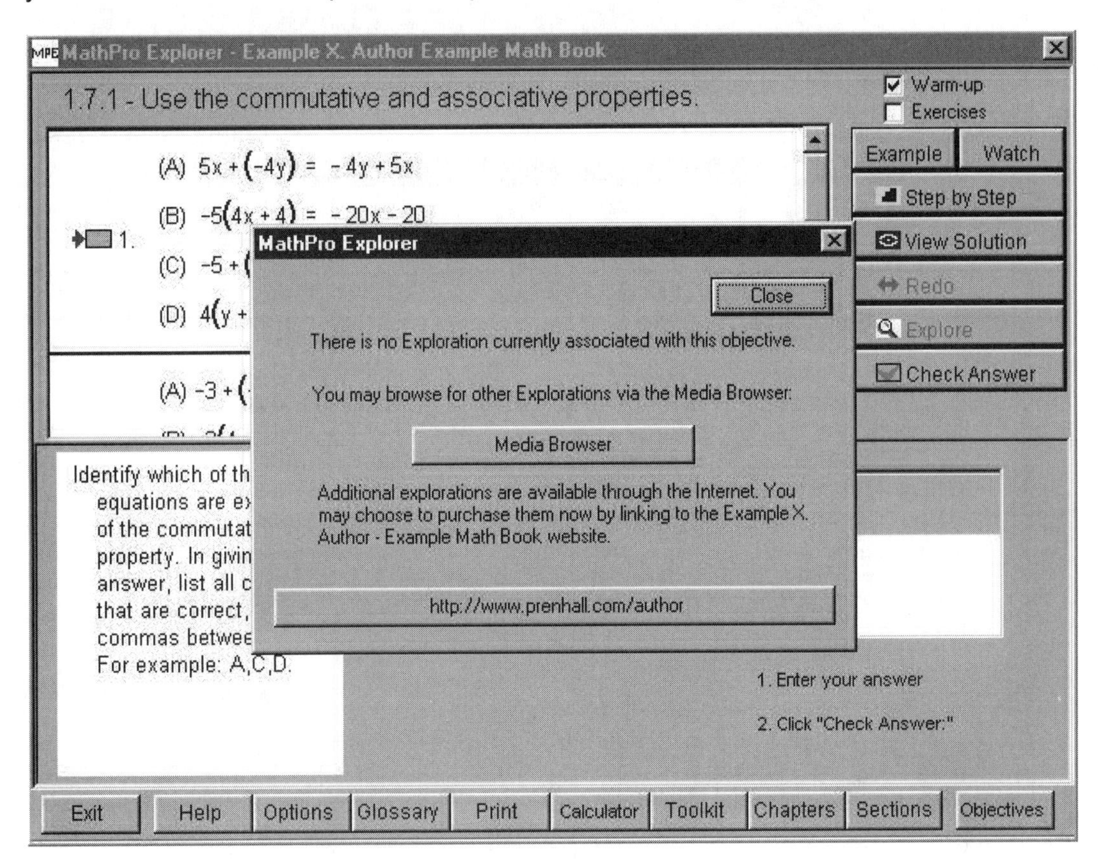

You can select the Media Browser to see a list of all existing Explorations for your book.

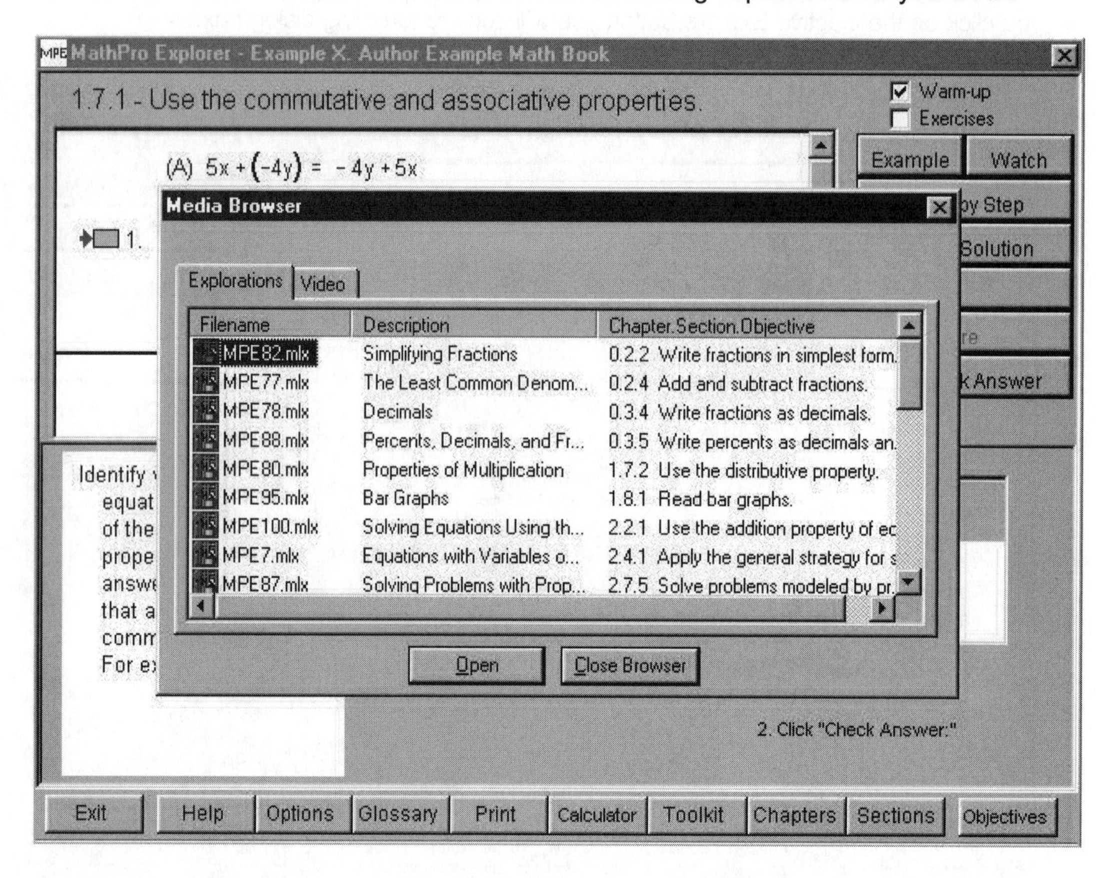

You can also launch the Media Browser at any time from the Media button on the main toolbar from the Chapter, Section or Objectives screens.

Chapters | Sections | Objectives The **Chapters, Sections** and **Objectives** buttons will close the problem window and return you to the selected location.

MathPro Explorer Toolkit

You can open Explorations from the Explore button or the Media Browser. In addition, you can click the Toolkit button on the main toolbar to access the Toolkit. Use the Toolkit to work any existing Explorations or to open your previously saved files.

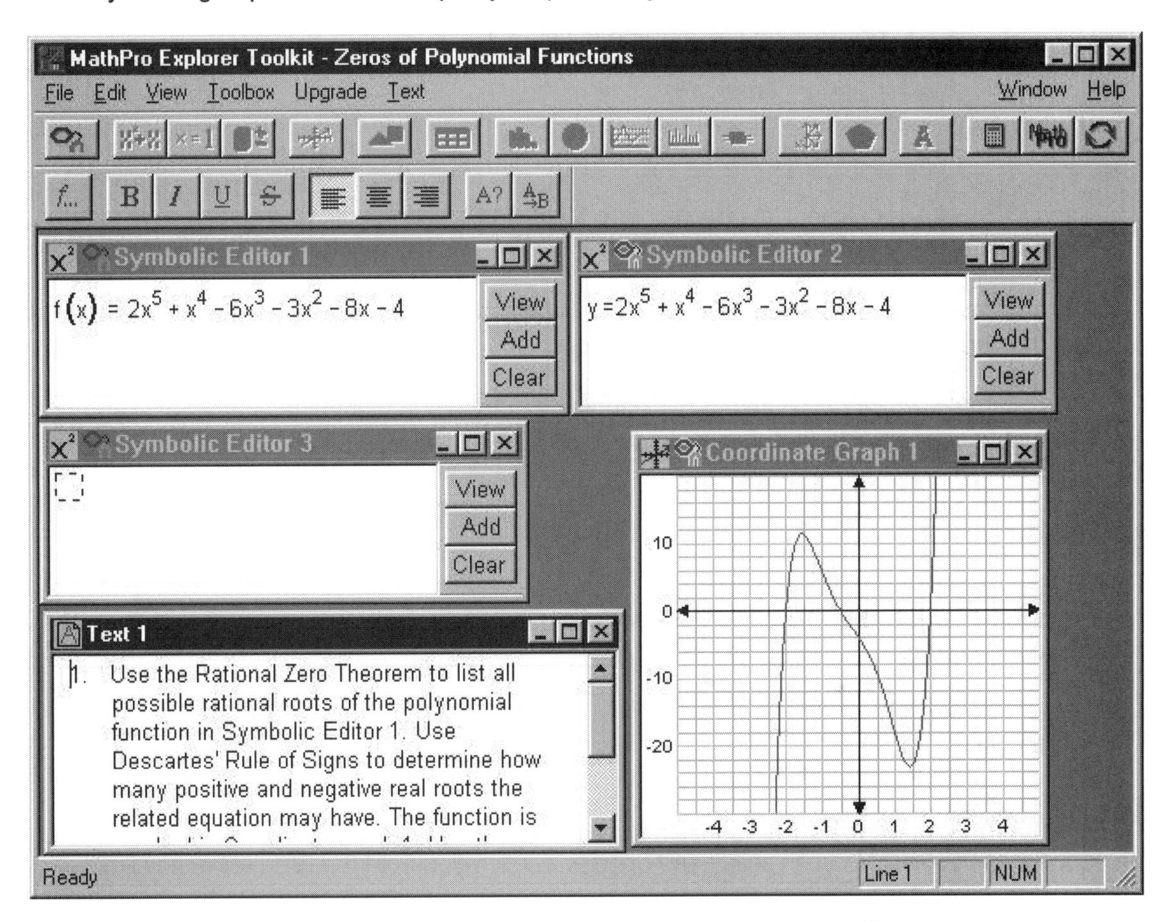

Explorer Button Definitions

Help

Help Click this button at any time to get Help.

Options

Options Click Options to change basic MathPro Explorer settings. You can save these settings as the default or use them for the current working session only.

Enable Quick Help - when checked, this will bring up small help boxes whenever you place the cursor over an active area on the screen.

Number of Problems Generated - Change the numbers to match the number of Warm-up and Exercise problems you want to generate.

Glossary

Glossary Use the Glossary to study the terminology used in the course or to look up a specific word.

Media

Media Click the Media button to open the Media Browser that lists all Explorations for this course. You can select and open an Exploration from the Browser at any time.

If your book contains Example Videos, click the Media button to open the Media Browser that lists all Example Videos for this book. You can select and open a Video from the Browser at any time.

Toolkit

Toolkit Click the Toolkit Button to access the Toolkit. Use the Toolkit to work any existing Explorations or to open your previously saved files.

Calculator

Calculator Click the Calculator button to bring up the system Calculator. This is helpful with many of the problems that require large calculations.

Summary

Summary You can view your performance by Section, Chapter, Date, or Book. Tabs allow you to select the level of information. A drop-down list in each tab enables you to select scores for different sections. Your scores are collected only on Exercise problems. If you have generated more than one set of Exercise problems, the scores from each set are saved.

Bookmark

Bookmark Use Bookmark when you see particular sections that you would like to revisit. Also, if you are interrupted while using MathPro Explorer, Bookmark will save your place. Click Set to place a bookmark in the section to which you want to return. Click Remove to delete the bookmark from a selected section. When you want to return to a bookmarked section, just reopen Bookmark and click the desired section.

Chapters

Chapters Click Chapters to bring up the Chapter screen that displays all the chapters in the book.

Sections

Sections Click Sections to bring up the Sections screen within a chapter.

Exit

Exit Leave MathPro Explorer at any time by clicking Exit. The system will ask you to confirm that you want to quit before the program exits.

Print

Print Click the Print button in the toolbar at the bottom of the Problem Sets to print out problem and status information for the current set.

Watch

Watch If your book contains Example Videos, click the Watch button to view a video of the author working a problem related to the current objective.

Check Answer

Check Answer Once you have entered an answer in the answer display, click Check Answer or press the Enter key on your keyboard to see if your answer is correct.

View Example/Example

View Example or Example When you have selected a problem, click the View Example or Example button to see the solution to a problem that is similar to the one you have selected. Use it as often as you like until you have mastered the process for solving that particular kind of problem.

Step by Step

Step by Step Click this button if you have difficulty solving the current problem. The Step by Step utility guides you through the process of solving any problem one step at a time. Use it to pinpoint the steps where you need help. Use it as often as you like within an objective until you have mastered the process for solving that particular kind of problem. You may only use Step by Step once per problem. If you use the Step by Step utility, you will not receive credit for the problem.

View Solution

View Solution Click this to view the solution to the current problem. If you use the View Solution utility, you will not receive credit for the problem.

Redo

Redo If you click Check Answer and your answer is incorrect, you have the option to go back and work the same type of problem again. Simply click on the incorrectly answered problem and click the Redo button. A new problem is generated. The system will track both the incorrect problem and the new problem, and the scores are added to the cumulative scores for the objective if you are working Exercise problems.

Explore

Explore If an Exploration accompanies the current objective, the Explore button is active. Open an Exploration by clicking the active button. When no Explorations appear for an objective, the Explore button is inactive. If you click on the inactive Explore button you will see a dialog box with a Media Browser link to the list of all existing Explorations for your book.

Symbol Buttons

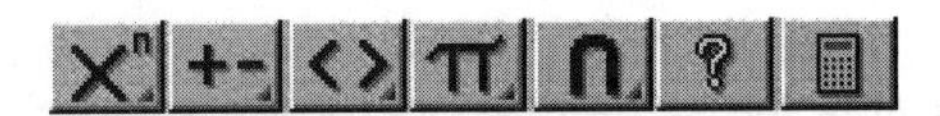

The small red arrow in the lower corner of the buttons indicates that there is a flip-up menu with additional buttons. Click any button to access its sub-menu. When a flip-up menu is showing, you can use the following for keyboard navigation:

Arrow keys - move up, down, right, left
Space bar (or Enter key) - selects the cell you have focus on
Esc key - closes the flip-up menu

Expressions

Click the Expressions button to access exponent, subscript, square root, nth root, rational, and absolute value symbols.

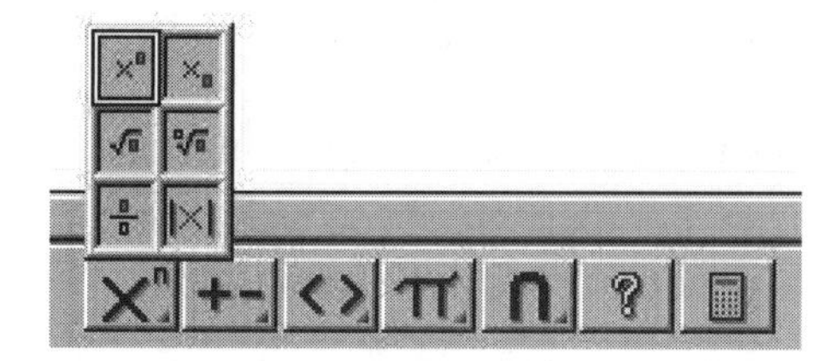

This icon enables you to enter a rational expression.

1) For example, to enter the fraction $\frac{2}{3}$, click the button, then type 2 in the top box, press the right arrow key, or click in the bottom box, and type 3.

2) If you have already typed an expression and want to make it the numerator of a rational expression, highlight the expression and click the button. For example, if you typed $3x^3 + 2y$ and you want to make it in the numerator of a rational expression, highlight the expression $3x^3 + 2y$ and click the button. This will give you the expression $\frac{3x^3 + 2y}{\square}$.

This icon enables you to enter an exponent.

1) For example, to enter x^2, type x and click the button. This gives you the expression $x\square$. Then type a 2, which will appear in the exponent box.

2) If you have already typed an expression and want to put it in an exponent, highlight the expression and click the button. For example, if you typed the expression $5y$ and you want to put it in an exponent, highlight the expression $5y$ and click the button. This will give you the expression $\square^{5y}$. You can click in the box or hit the left arrow key to enter a base.

3) When you are done typing an exponent on a base, remember to click your cursor to the right of the exponential expression to leave the superscript mode.

This icon enables you to enter subscripts.

1) For example, to enter x_1, type x and click the button. This will give you $x\,\square$. Then type a 1 in the subscript box.

2) If you have already typed an expression and want to put it in a subscript, highlight the expression and click the button. For example, if you typed the expression xy and you want to put it in a subscript, highlight the expression xy and click the button. This will give you the expression $\square_{y}$. You can click in the box or hit the left arrow key to enter a base.

3) When you are done typing a subscript on a base, remember to click your cursor to the right of the expression to leave the subscript mode.

This icon enables you to enter a square root.

1) For example, to enter $\sqrt{x+9}$ click the button and type $x + 9$ in the radicand box.

2) If you have already entered an expression and want to put it in a radical, highlight the expression and click the button. For example if you typed $\dfrac{-b \pm b^2 - 4ac}{2a}$ and you wanted to put the expression $b^2 - 4ac$ under a

radical, highlight just that part of the expression and click the button. This will give you the expression $\frac{-b \pm \sqrt{b^2 - 4ac}}{2a}$.

This icon enables you to enter any root.

1) For example, to enter $\sqrt[4]{16}$, click the button. The cursor will be in the index (smaller) box outside the radical. Type 4. Then click in the box under the radical or press the right arrow key, and type 16.

2) If you have already typed an expression, you can highlight the expression and click the button. This will put the expression under the radical and put the cursor in the small box outside the radical.

This icon enables you to enter absolute value expressions.

Binary Operators

Click the Binary Operators button to access addition, subtraction, multiplication, division, and plus/minus symbols.

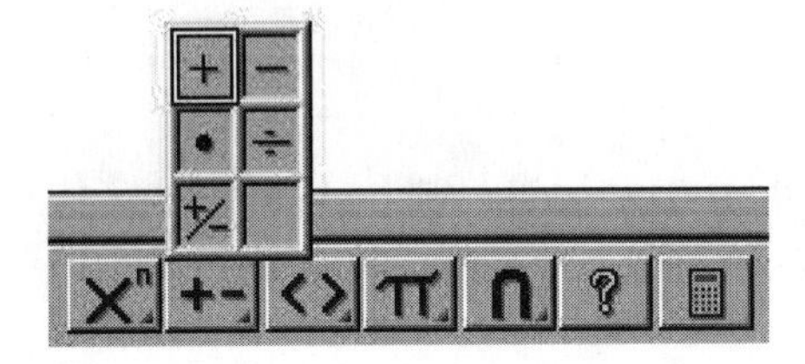

+ Inserts a plus sign at the position of the cursor.

− Inserts a minus sign at the position of the cursor.

• Inserts a multiplication sign at the position of the cursor.

÷ Inserts a division sign at the position of the cursor.

± Inserts a plus/minus sign at the position of the cursor.

Relational Operators

Click the Relational Operators button to access less than, less than or equal to, greater than, greater than or equal to, equal, or not equal symbols.

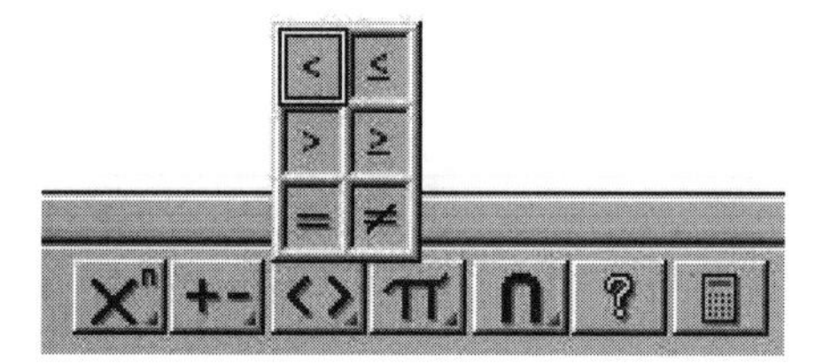

Inserts a less than symbol at the position of the cursor.

Inserts a less than or equal to symbol at the position of the cursor.

Inserts a greater than symbol at the position of the cursor.

Inserts a greater than or equal to symbol at the position of the cursor.

Inserts a equal to symbol at the position of the cursor.

Inserts a not equal symbol at the position of the cursor.

Constants

Click the Constants button to access constant *e*, constant *i*, pi, or infinity symbols.

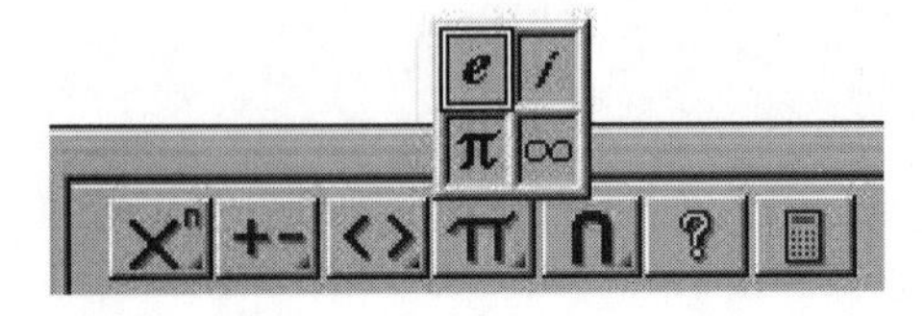

Inserts a constant *e* symbol at the position of the cursor.

Inserts a constant *i* symbol at the position of the cursor.

Inserts a Pi symbol at the position of the cursor.

Inserts an infinity symbol at the position of the cursor.

Set Operators

Click the Set Operators button to access the set union, set intersection, member of set and not member of set symbols.

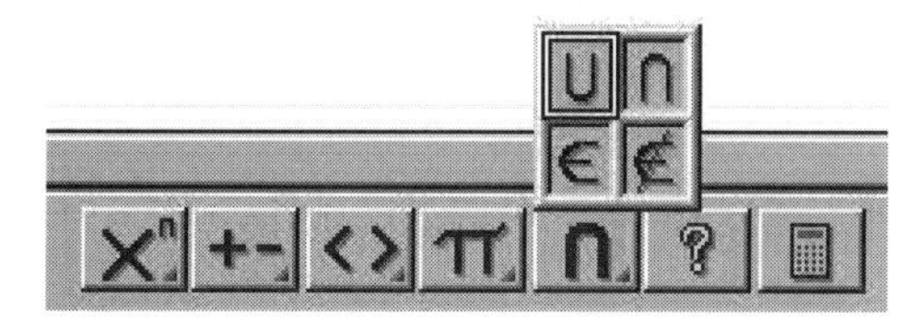

∪ Inserts a set union symbol at the position of the cursor.

∩ Inserts a set intersection symbol at the position of the cursor.

∈ Inserts a member of set symbol at the position of the cursor.

∉ Inserts a not member of set symbol at the position of the cursor.

Special Buttons

? Click the Help icon button to open the Entering Math Expressions screen of the Help file.

Click the Calculator button to bring up the system Calculator. This is helpful with many of the problems that require large calculations. You can switch between the standard and the scientific calculator by clicking View in the menu bar and choosing the calculator you want.

Entering Math Expressions

You can enter math symbols and build mathematical expressions and equations by using the keyboard or the buttons on the Math Editor toolbar submenus.

Using the Symbol Buttons

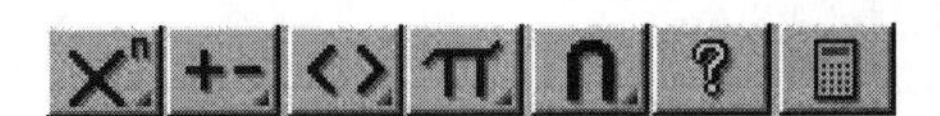

There are two methods for using the buttons.

1) Click the button to access the submenu for the desired expression. Select the specific symbol needed. That expression will be inserted in the Math Editor Window at the position of the blinking cursor. Use the mouse to move the cursor to the desired dotted rectangle then type your answer.

   Use the mouse or the right and left arrow keys to move the position of the cursor.

2) If there is red on a button in a pop-up submenu, you can highlight an expression in the Math Editor Window and click the desired button of the symbol you want. This will insert that expression where you see the red in the Icon box.

Using the Keyboard

In the lists that follow, note that "Ctrl-" indicates you must hold down the Control (Ctrl) key while pressing the following key, as summarized below.

Special/Editing Keys

Ctrl-Z: Undo the most recent operation (also, Alt-Backspace)
Ctrl-X: Cuts a section out of the expression
Ctrl-C: Copies a section
Ctrl-V: Pastes the most recent cut or copy block
Ctrl-Y: Redo the most recent operation (also, Shift-Alt-Backspace)
Home: Start of equation
Left arrow: Left
Right arrow: Right
End: End of equation

Selection Keys

Shift-Home: Selects all text from cursor position to start of text
Shift-Left arrow: Selects character to left of cursor

Shift-Right arrow: Selects character to right of cursor
Shift-End: Selects all text from cursor position to end of text

Operation Symbols

Less Than Equal: Ctrl-,
Greater Than Equal: Ctrl-.
Not Equal: Ctrl-3
Pi: Ctrl-p
Plus-Minus: Ctrl-=
Divide: Ctrl-:
Multiply: Ctrl-8

Special Structures

Rational: Ctrl-/
Square Root: Ctrl-s
Root: Ctrl-r
Exponent: Ctrl-6
Subscript: Ctrl--
Absolute: Ctrl-A
Summation: Ctrl-U
mxn Matrix: Ctrl-M
2x2 Matrix: Ctrl-2
Vector: Ctrl-1
Constant *e*: Ctrl-E
Constant *i*: Ctrl-I
Log: Ctrl-L
Ln: Ctrl-N
Log, base *n*: Ctrl-B
Inverse Log: Ctrl-Alt-L
Inverse Ln: Ctrl-Alt-N
Sin: Ctrl-H
Cos: Ctrl-O
Tan: Ctrl-T
Inverse Sin: Ctrl-Alt-H
Inverse Cos: Ctrl-Alt-O
Inverse Tan: Ctrl-Alt-T
Permutation: Ctrl-Alt-P
Combination: Ctrl-Alt-C
Union: Ctrl-9
Intersection: Ctrl-0

MathPro Explorer 4.0 Home/Single User
Administrator Program

MathPro Explorer Administrator Overview

The MathPro Explorer Administrator Program provides you with several different options. You can generate reports or retrieve saved reports, which include results from Exercise problems and Practice Tests. In addition, you can create Practice Tests to take in MathPro Explorer. The On-Line Help feature is available for convenient assistance.

Administrator Login

Upon entering the MathPro Explorer Administrator Program, you are prompted to enter a password. The default is no password; therefore, you should set the password the first time you enter the Administrator Program. Click OK on the Login screen to open the MathPro Explorer Administrator main menu.

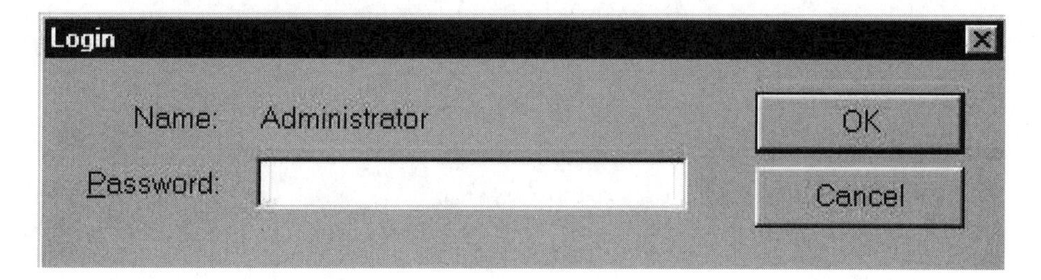

MathPro Explorer Administrator Main Menu

The MathPro Explorer Administrator main menu appears at the top of the screen below. You have the option to use the main menu or the toolbar buttons to navigate through the program.

Toolbar Buttons

Click the New Report button to generate a Summary Report or Practice Test Report.

Click the Open Report button to open a previously saved Summary Report or Practice Test Report.

Click the Save Report button to save a current Summary Report or Practice Test Report.

Click the Print button to print an open Summary Report or Practice Test Report.

Click the Refresh button to refresh an open Summary Report if students are working Exercise problems in MathPro Explorer.

Click the Help button to access the on-line Administrator help menu at any time.

Creating Reports

To generate a report, select File from the main menu, then select New Report as displayed below or click the New Report button.

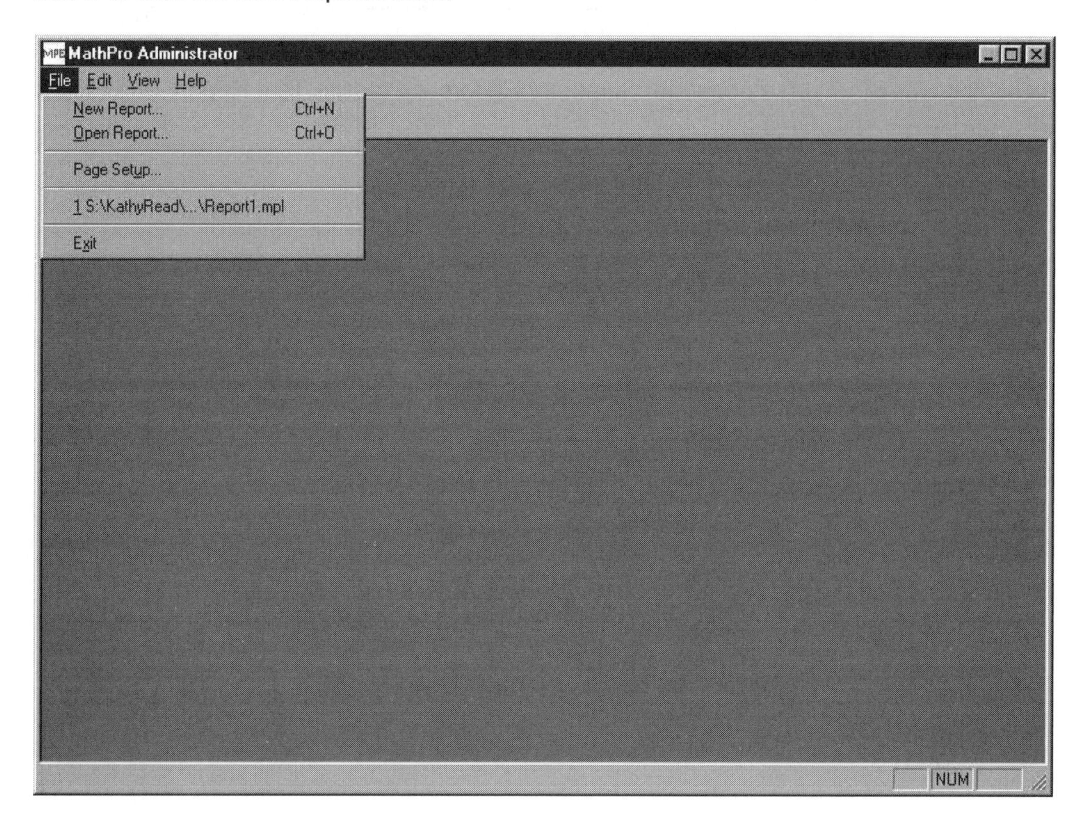

Once New Report is selected, the following dialog box opens.

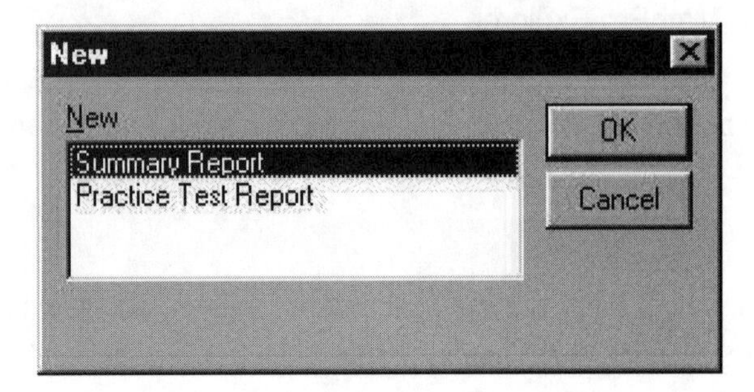

Select the type of report to generate, Summary Report or Practice Test Report, and click OK.

Summary Report

Type the desired Chapter information in the corresponding field or use the list box to select the desired Chapter information for the report. Once you enter this information, use your mouse to check the desired data to include (Book, Chapter, Section, Objective, Cumulative, or By Date) and click the View Report button.

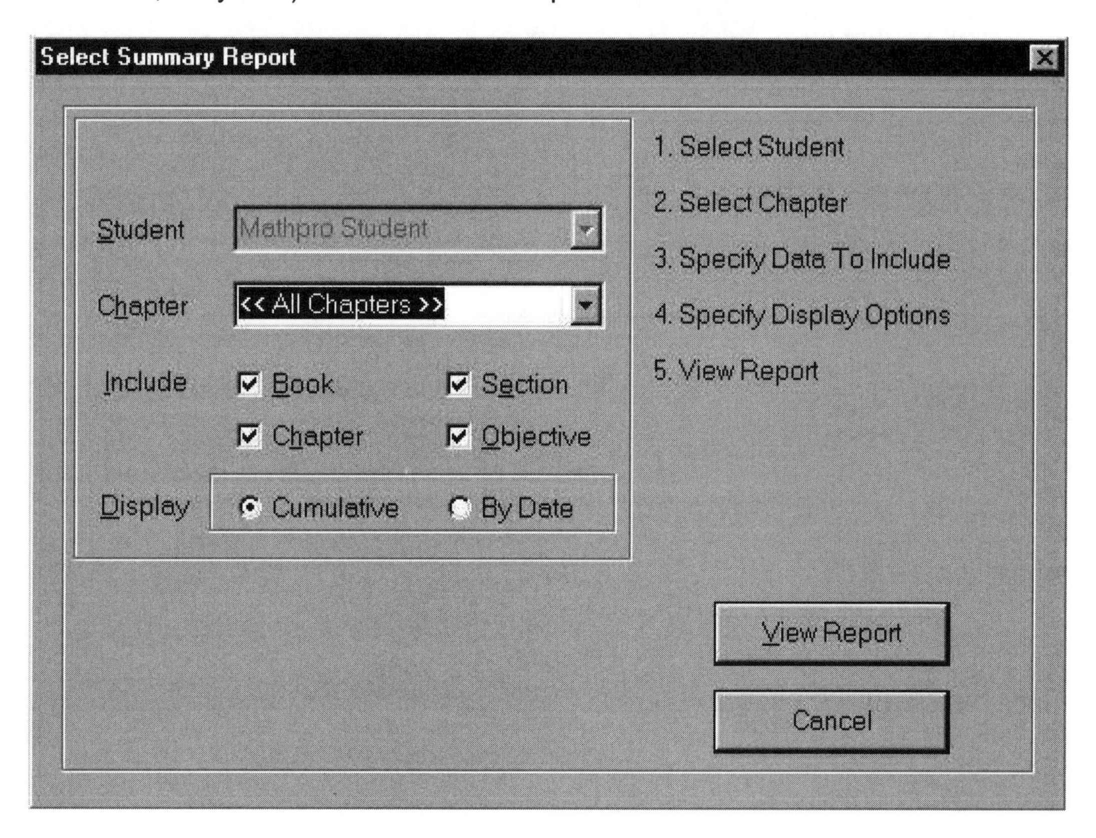

Upon selecting the View Report button, the Administrator Program generates a report similar to the one shown below. (The report is a text file that you can view, save and reopen, or print.)

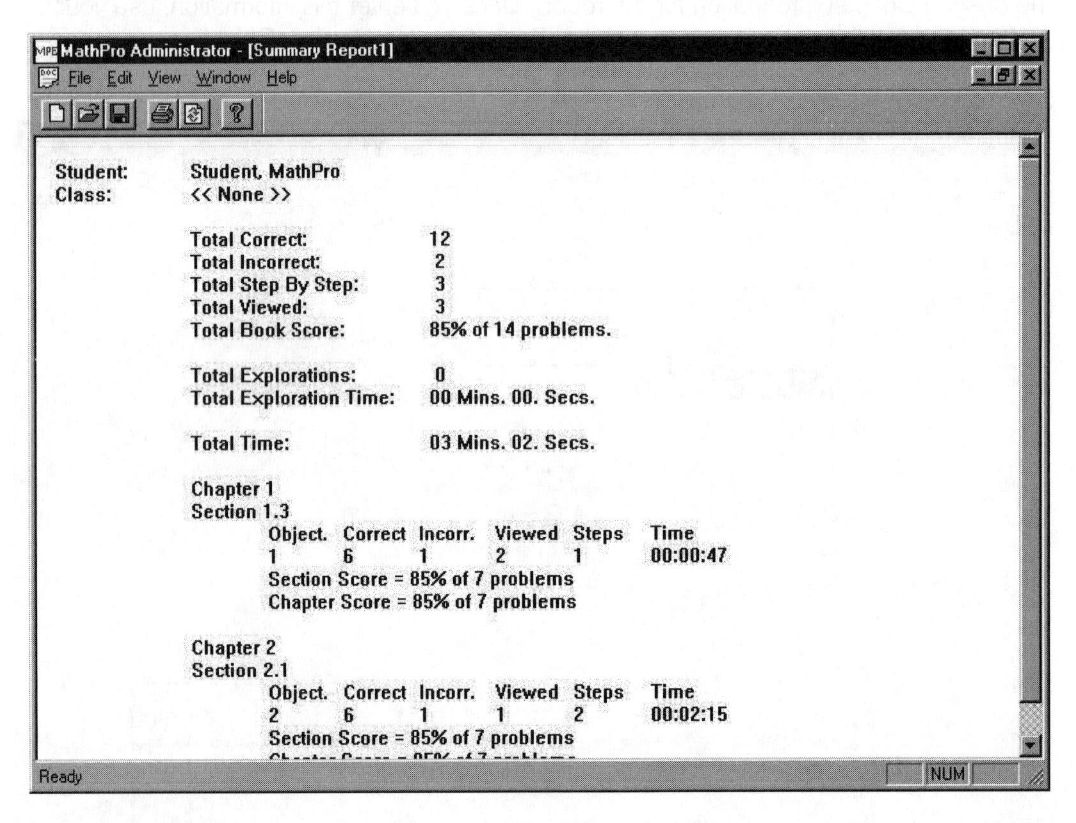

Practice Test Report

Type the desired Chapter information in the corresponding field or use the list box to select the desired Chapter information for the report. Once you enter this information, click the View Report button.

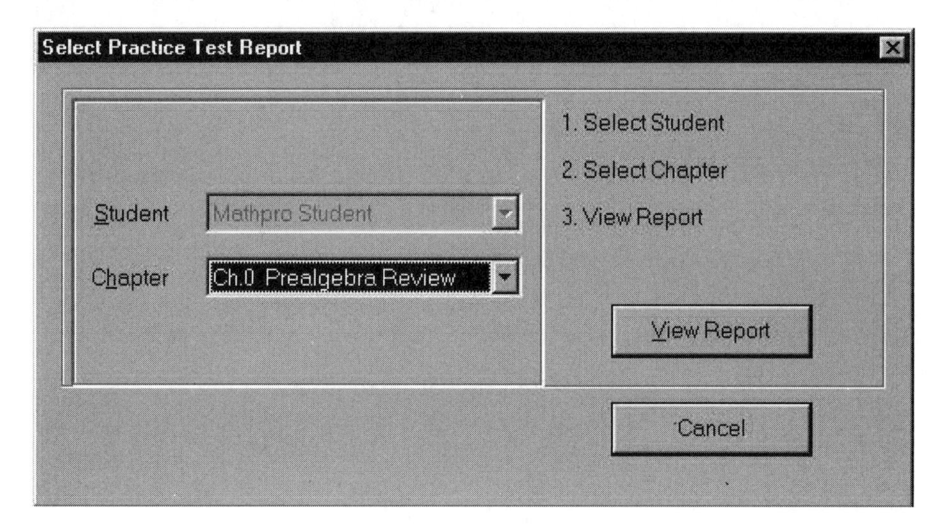

Upon selecting the View Report button, the Administrator Program generates a report similar to the one shown below. (The report is a text file that you can view, save and reopen, or print.)

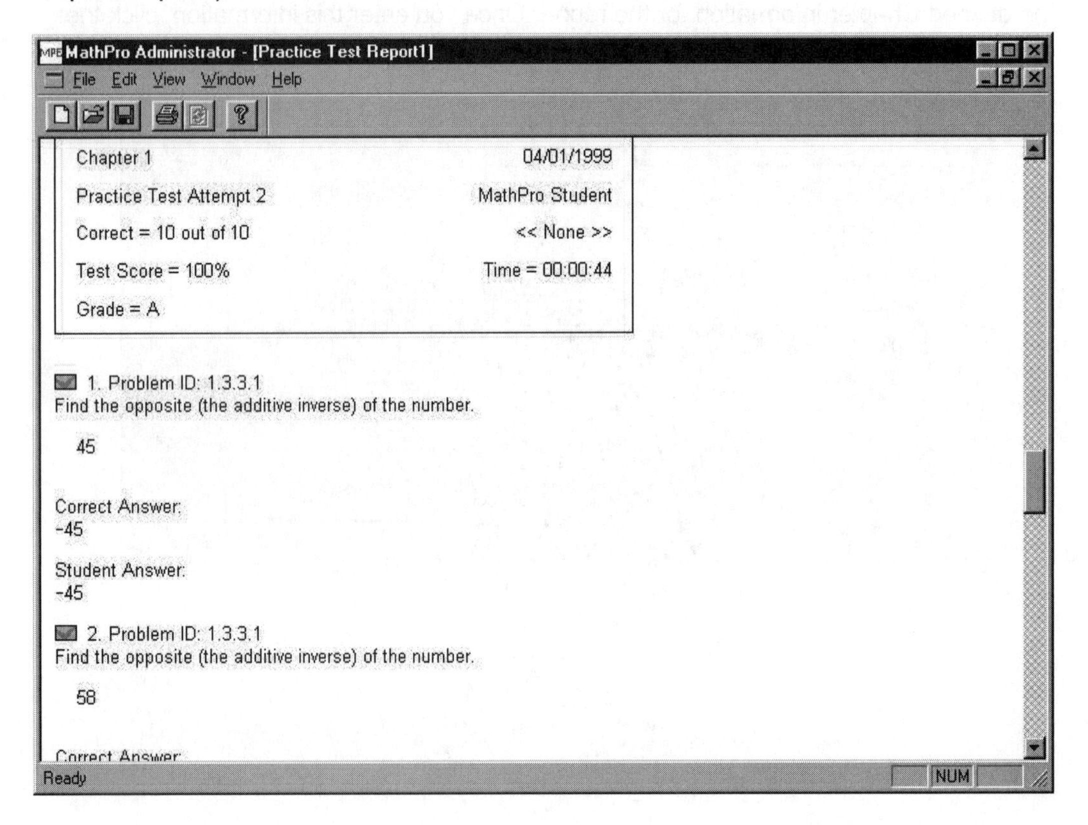

Creating Practice Tests and Changing the Password

From the MathPro Explorer Administrator Program, you have the option to create Practice Tests or Change Password. You can choose these options from the Edit menu.

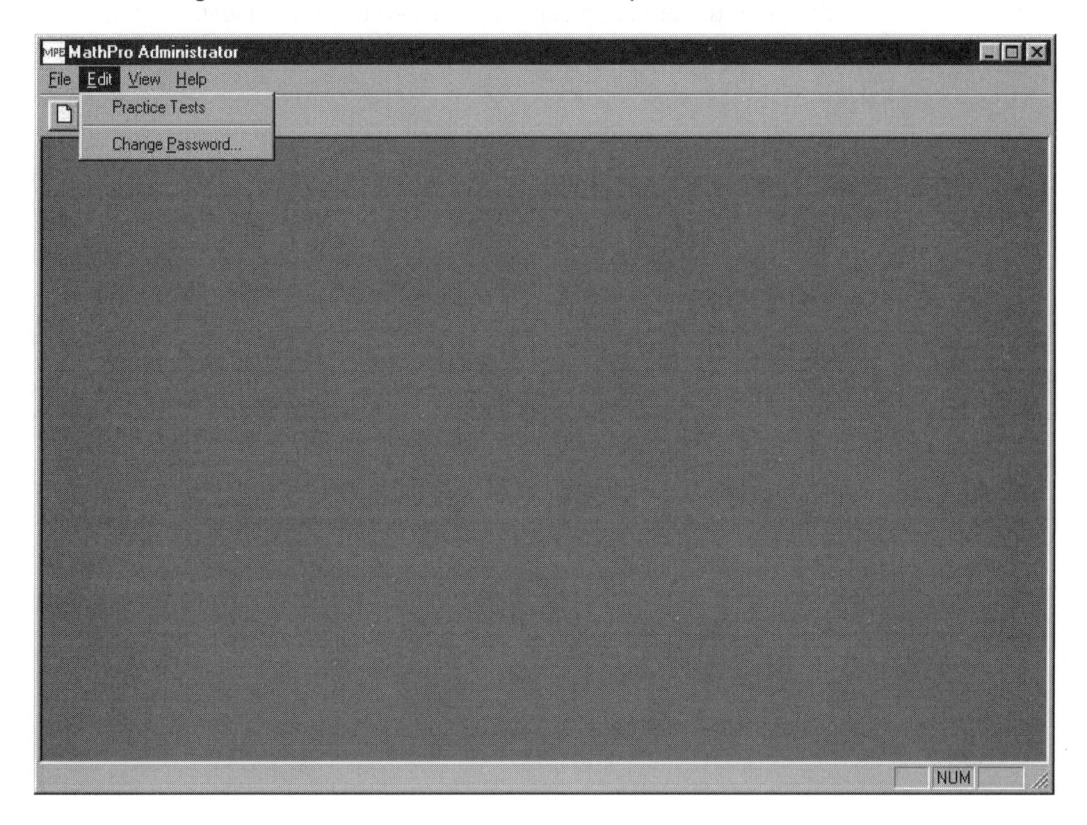

Creating Practice Tests

Select Practice Tests from the Edit menu to open the Practice Chapter Tests dialog box. Through this screen, you can enter the chapter, the number of problems, the maximum number of tries and the minimum grade percentages desired to create a test.

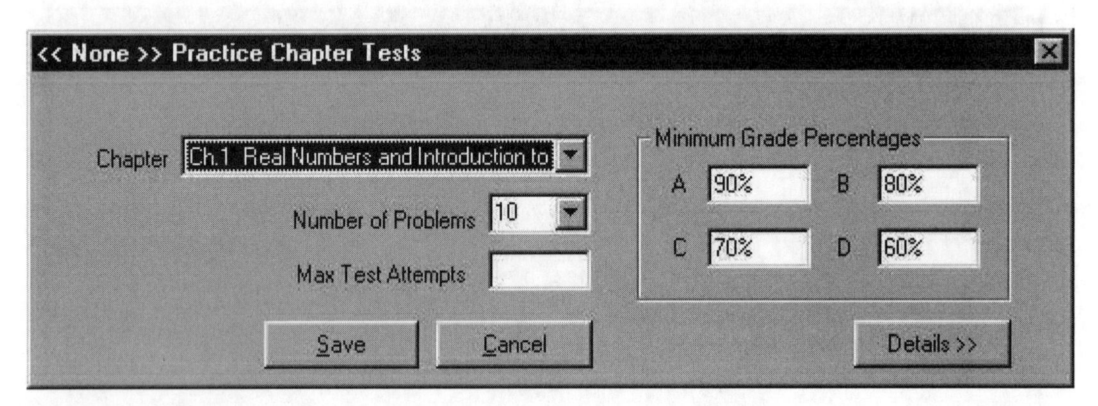

Click the Details button to view the chapter problems. Use the mouse to navigate through the tree structure on the left side of the dialog box under Chapter Problems, highlight a problem, and view a Sample Problem in the lower portion of the screen. The Add button adds a highlighted problem to the Selected Problem list on the right side of the dialog box. The Fill button creates a randomly generated set of problems from the chapter, dependent upon the Number of Problems selected. Once problems have been selected, you have the option to remove one problem or remove all problems by clicking on either the Remove or Remove All button, respectively.

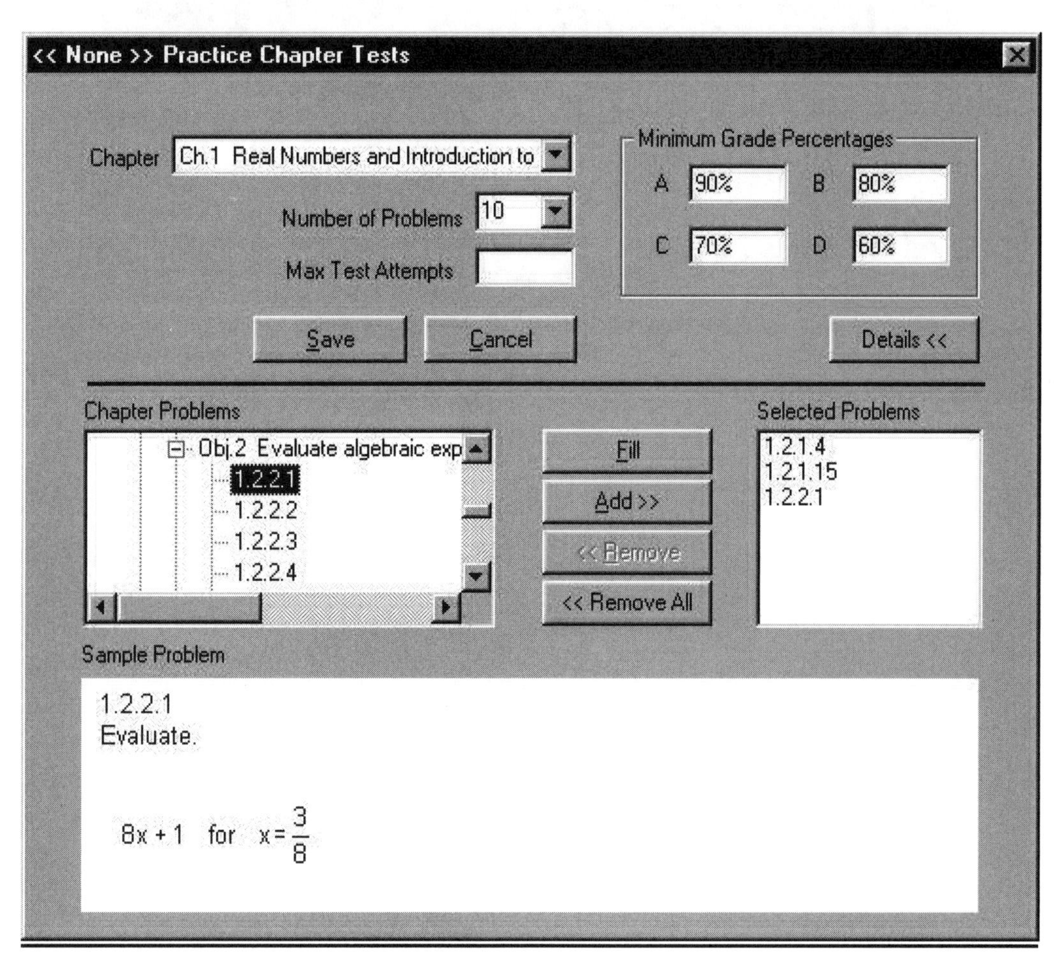

Changing the Password

Select Change Password from the Edit menu to open the Change Password dialog box. Type a new password in the New Password field and repeat it in the Confirm Password field. Click the OK button to initiate the change or click Cancel to abort the change. Once you change the password, you must remember the new password to log back into the MathPro Explorer Administrator Program.

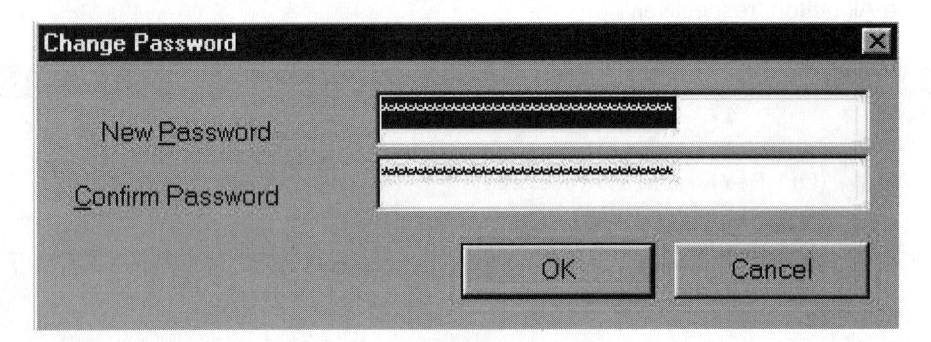

Record-Keeping Information

Section Records

Student information is collected on Exercise problems only. After you work Exercise problems in an objective, the Administrator Program saves a record of your activities.

Click the Summary button from the Chapter, Section, or Objective screens to open the Scores For <User> dialog box. The Section Scores information gives you feedback on the work completed in each section. The information includes:

- A summary of all objectives worked within a section
- The number of Exercise problems solved correctly or incorrectly for each objective
- The number of times View Solution and Step by Step have been accessed
- The length of time spent on Exercises for each objective
- The names of any Explorations worked for each objective
- The amount of time spent on each Exploration
- A cumulative total for each of these categories, as well as a cumulative score percentage for the section.

Click the Print button for a printed report of the Section Score information.

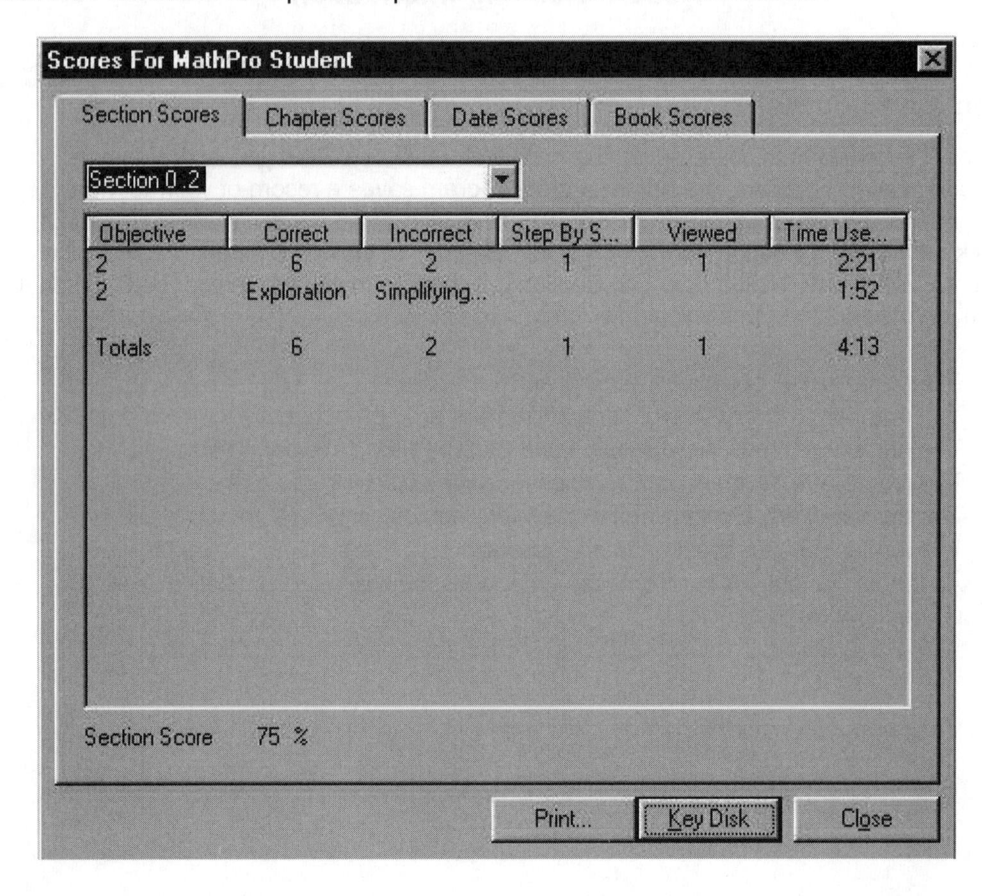

Chapter Records

The Chapter Scores information provides:

- A summary of performance by chapter
- A summary of all sections worked within each chapter
- The number of Exercise problems solved correctly or incorrectly for each section
- The number of times View Solution and Step-by-Step have been accessed
- The number of Explorations worked for each section
- The amount of time spent on each section
- A cumulative total for each of these categories, as well as a cumulative score percentage for the chapter.

Click the Print button for a printed report of the Chapter Score information.

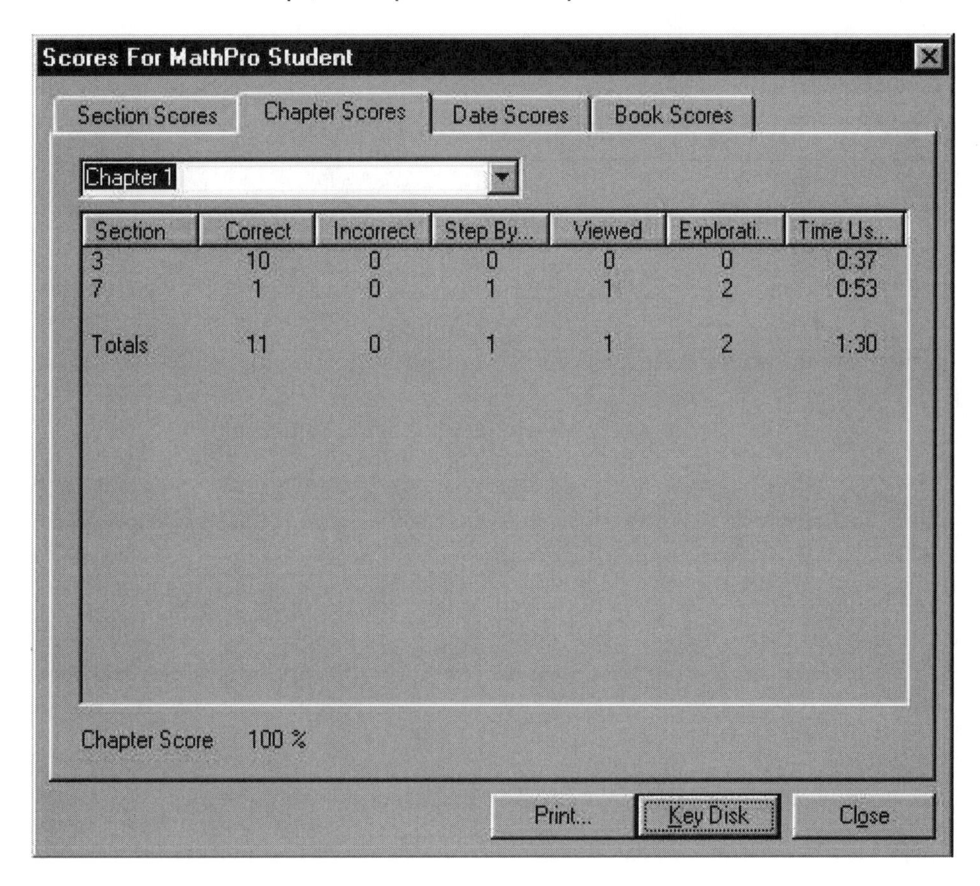

Date Records

The Date Scores information provides:

- A summary of all objectives worked within a section on a specific date
- The number of Exercise problems solved correctly or incorrectly for each objective
- The number of times View Solution and Step by Step have been accessed
- The length of time spent on Exercises for each objective
- The names of any Explorations worked for each objective
- The amount of time spent on each Exploration
- A cumulative total for each of these categories, as well as a cumulative score percentage for the section.

Click the Print button for a printed report of the Date Score information.

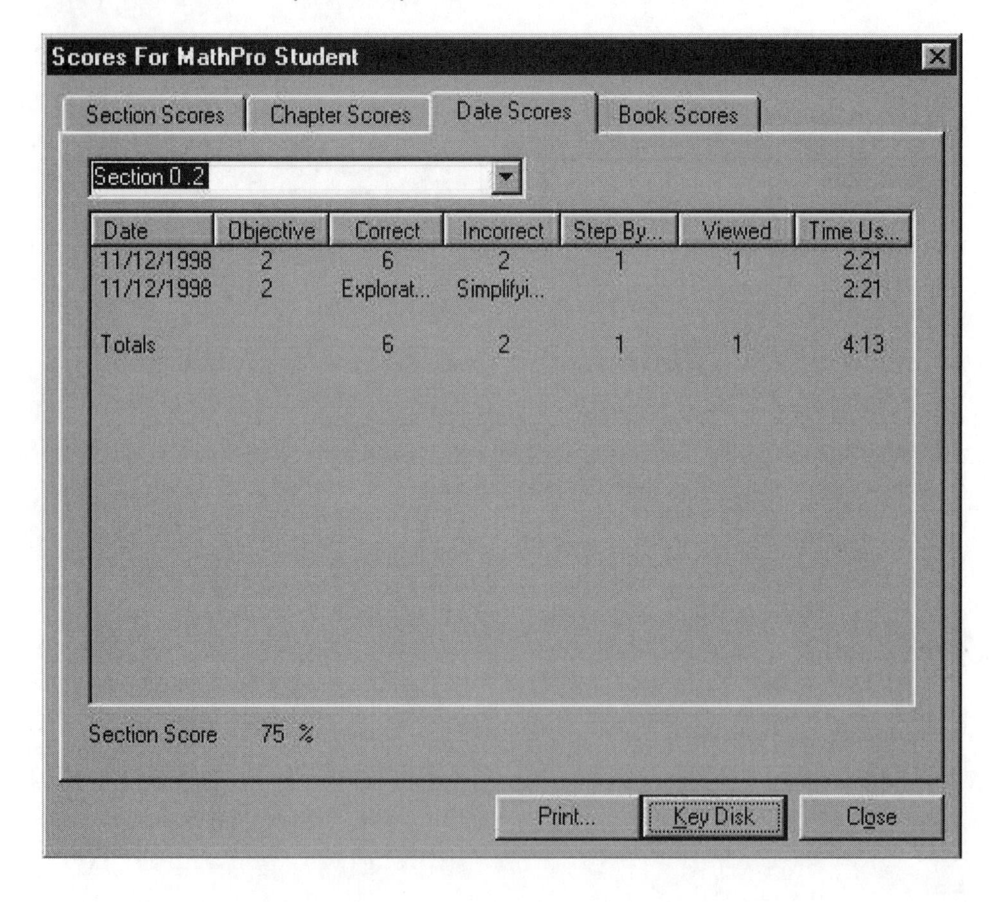

Book Records

The Book Scores information provides:

- The book totals of Exercise problems solved correctly and incorrectly
- The number of times View Solution and Step by Step have been accessed
- The number of Explorations worked
- The total amount of time spent on Explorations
- The total number of Exercise problems attempted in the book
- The percentage of correct answers
- The amount of time spent on Exercise problems
- The number of sessions and the total time spent working Exercise problems in MathPro Explorer.

Click the Print button for a printed report of the Book Score information.

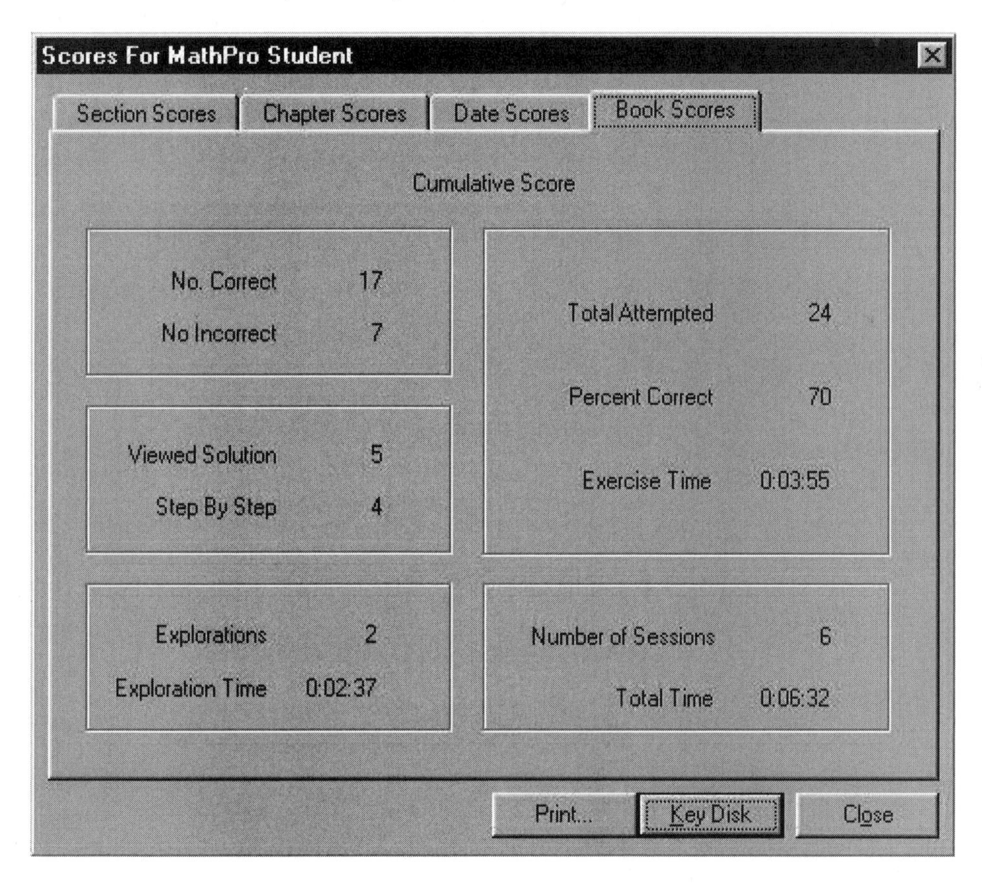

MathPro Development Team

Programming and Design

Ian Seekell	Jeff Nassiff

Content

Philip Lanza	Jack Janssen
Sheri Michel	Michael Duguay
Katherine Gregory	Brian Muller

SINGLE PC LICENSE AGREEMENT AND LIMITED WARRANTY

READ THIS LICENSE CAREFULLY BEFORE OPENING THIS PACKAGE. BY OPENING THIS PACKAGE, YOU ARE AGREEING TO THE TERMS AND CONDITIONS OF THIS LICENSE. IF YOU DO NOT AGREE, DO NOT OPEN THE PACKAGE. PROMPTLY RETURN THE UNOPENED PACKAGE AND ALL ACCOMPANYING ITEMS TO THE PLACE YOU OBTAINED THEM [[FOR A FULL REFUND OF ANY SUMS YOU HAVE PAID FOR THE SOFTWARE]]. THESE TERMS APPLY TO ALL LICENSED SOFTWARE ON THE DISK EXCEPT THAT THE TERMS FOR USE OF ANY SHAREWARE OR FREEWARE ON THE DISKETTES ARE AS SET FORTH IN THE ELECTRONIC LICENSE LOCATED ON THE DISK:

1. GRANT OF LICENSE and OWNERSHIP: The enclosed computer programs and data ("Software") are licensed, not sold, to you by Prentice-Hall, Inc. ("We" or the "Company") and in consideration of your purchase or adoption of the accompanying Company textbooks and/or other materials, and your agreement to these terms. We reserve any rights not granted to you. You own only the disk(s) but we and/or our licensors own the Software itself. This license allows you to use and display your copy of the Software on a single computer (i.e., with a single CPU) at a single location for academic use only, so long as you comply with the terms of this Agreement. You may make one copy for back up, or transfer your copy to another CPU, provided that the Software is usable on only one computer.

2. RESTRICTIONS: You may not transfer or distribute the Software or documentation to anyone else. Except for backup, you may not copy the documentation or the Software. You may not network the Software or otherwise use it on more than one computer or computer terminal at the same time. You may not reverse engineer, disassemble, decompile, modify, adapt, translate, or create derivative works based on the Software or the Documentation. You may be held legally responsible for any copying or copyright infringement which is caused by your failure to abide by the terms of these restrictions.

3. TERMINATION: This license is effective until terminated. This license will terminate automatically without notice from the Company if you fail to comply with any provisions or limitations of this license. Upon termination, you shall destroy the Documentation and all copies of the Software. All provisions of this Agreement as to limitation and disclaimer of warranties, limitation of liability, remedies or damages, and our ownership rights shall survive termination.

4. LIMITED WARRANTY AND DISCLAIMER OF WARRANTY: Company warrants that for a period of 60 days from the date you purchase this SOFTWARE (or purchase or adopt the accompanying textbook), the Software, when properly installed and used in accordance with the Documentation, will operate in substantial conformity with the description of the Software set forth in the Documentation, and that for a period of 30 days the disk(s) on which the Software is delivered shall be free from defects in materials and workmanship under normal use. The Company does not warrant that the Software will meet your requirements or that the operation of the Software will be uninterrupted or error-free. Your only remedy and the Company's only obligation under these limited warranties is, at the Company's option, return of the disk for a refund of any amounts paid for it by you or replacement of the disk. THIS LIMITED WARRANTY IS THE ONLY WARRANTY PROVIDED BY THE COMPANY AND ITS LICENSORS, AND THE COMPANY AND ITS LICENSORS DISCLAIM ALL OTHER WARRANTIES, EXPRESS OR IMPLIED, INCLUDING WITHOUT LIMITATION, THE IMPLIED WARRANTIES OF MERCHANTABILITY AND FITNESS FOR A PARTICULAR PURPOSE. THE COMPANY DOES NOT WARRANT, GUARANTEE OR MAKE ANY REPRESENTATION REGARDING THE ACCURACY, RELIABILITY, CURRENTNESS, USE, OR RESULTS OF USE, OF THE SOFTWARE.

5. LIMITATION OF REMEDIES AND DAMAGES: IN NO EVENT, SHALL THE COMPANY OR ITS EMPLOYEES, AGENTS, LICENSORS, OR CONTRACTORS BE LIABLE FOR ANY INCIDENTAL, INDIRECT, SPECIAL, OR CONSEQUENTIAL DAMAGES ARISING OUT OF OR IN CONNECTION WITH THIS LICENSE OR THE SOFTWARE, INCLUDING FOR LOSS OF USE, LOSS OF DATA, LOSS OF INCOME OR PROFIT, OR OTHER LOSSES, SUSTAINED AS A RESULT OF INJURY TO ANY PERSON, OR LOSS OF OR DAMAGE TO PROPERTY, OR CLAIMS OF THIRD PARTIES, EVEN IF THE COMPANY OR AN AUTHORIZED REPRESENTATIVE OF THE COMPANY HAS BEEN ADVISED OF THE POSSIBILITY OF SUCH DAMAGES. IN NO EVENT SHALL THE LIABILITY OF THE COMPANY FOR DAMAGES WITH RESPECT TO THE SOFTWARE EXCEED THE AMOUNTS ACTUALLY PAID BY YOU, IF ANY, FOR THE SOFTWARE OR THE ACCOMPANYING TEXTBOOK. BECAUSE SOME JURISDICTIONS DO NOT ALLOW THE LIMITATION OF LIABILITY IN CERTAIN CIRCUMSTANCES, THE ABOVE LIMITATIONS MAY NOT ALWAYS APPLY TO YOU.

6. GENERAL: THIS AGREEMENT SHALL BE CONSTRUED IN ACCORDANCE WITH THE LAWS OF THE UNITED STATES OF AMERICA AND THE STATE OF NEW YORK, APPLICABLE TO CONTRACTS MADE IN NEW YORK, AND

SHALL BENEFIT THE COMPANY, ITS AFFILIATES AND ASSIGNEES. HIS AGREEMENT IS THE COMPLETE AND EXCLUSIVE STATEMENT OF THE AGREEMENT BETWEEN YOU AND THE COMPANY AND SUPERSEDES ALL PROPOSALS OR PRIOR AGREEMENTS, ORAL, OR WRITTEN, AND ANY OTHER COMMUNICATIONS BETWEEN YOU AND THE COMPANY OR ANY REPRESENTATIVE OF THE COMPANY RELATING TO THE SUBJECT MATTER OF THIS AGREEMENT. If you are a U.S. Government user, this Software is licensed with "restricted rights" as set forth in subparagraphs (a)-(d) of the Commercial Computer-Restricted Rights clause at FAR 52.227-19 or in subparagraphs (c)(1)(ii) of the Rights in Technical Data and Computer Software clause at DFARS 252.227-7013, and similar clauses, as applicable.

Should you have any questions concerning this agreement or if you wish to contact the Company for any reason, please contact in writing: New Media / Higher Education Division / Prentice Hall Inc. / 1 Lake Street / Upper Saddle River, NJ 07458.